Molecular Genetics of Stroke

Colloquium Series on Genomic and Molecular Medicine

Editor
Dhavendra Kumar
The University of Glamorgan
UK Institute of Medical Genetics
Cardiff University School of Medicine
University Hospital of Wales

The progress of medicine has always been driven by advances in science and technology. The practice of medicine at a given place and time is a reflection of the current knowledge, applications of the available information and evidence, social/cultural/ religious beliefs, and statutory requirements. Thus, it needs to be dynamic and flexible to accommodate changes and new developments in basic and applied science in keeping with the individual and societal expectations. From 1970 onwards, there has been a continuous and growing recognition of the molecular basis of medical practice. Most medical curricula allow sufficient space and time to ensure satisfactory coverage of the basic principles of molecular biology. Emphasis is given on the relevance of molecular science in the practice of clinical medicine paving the way for a more holistic approach to patient care utilizing new dimensions in diagnostics and therapeutics. It is extremely important that both teachers and students have an agreed agenda for learning and applications of molecular medicine. This should not be restricted to few uncommon genetic conditions but extended to include inflammatory conditions, infectious diseases, cancer, and age-related degenerative conditions involving multiple body systems.

Alongside the developments and progress in molecular medicine, rapid and new discoveries in genetics led to an entirely new approach to the practice of clinical medicine. However, the field of genetic medicine has been restricted to the diagnosis, offering explanation and assistance to patients and clinicians in dealing with a number of relatively uncommon inherited disorders. Nevertheless, this field has gradually established and accorded the specialty status in the medical curriculum of several countries.

Since the completion of the human genome in 2003 and several other genomes, there is now a plethora of information available that has attracted the attention of molecular biologists and allied researchers. A new biological science of genomics is now with us with far reaching dimensions and applications. Rapid and revealing findings in genomics have led to changes in the fundamental concepts in cell and molecular biology. A present day student of biology is expected to conceptualize the sequence of genome-gene-molecule-cell with reference to specific tissue, organ, and a body system. In other words, evolutionary and morbid changes at the genome level could be the basis of normal human variation and disease. During the last decade, rapid progress has been made in new genome-level diagnostic and prognostic laboratory methods. Applications of individual genomic information in clinical medicine have led to the prospect of robust evidence-based personalized medicine. Genomics has led to the discovery and development of a number of new drugs with far-reaching implications in pharmaco-therapeutics. The existence of Genomic Medicine around us is inseparable from molecular medicine. Both genomic and molecular medicines are in fact two dimensions of the integrated modern molecular medicine with tremendous implications for the future of clinical medicine.

Molecular Genetics of Stroke
Yoshiji Yamada
www.morganclaypool.com

ISBN: 9781615043705 paperback

ISBN: 9781615043712 ebook

DOI: 10.4199/C00052ED1V01Y201204GMM001

A Publication in the

COLLOQUIUM SERIES ON GENOMIC AND MOLECULAR MEDICINE

Lecture #1

Series Editor: Dhavendra Kumar, Cardiff University School of Medicine, University Hospital of Wales

Series ISSN Pending

Molecular Genetics of Stroke

Yoshiji Yamada
Department of Human Functional Genomics
Life Science Research Center
Mie University

COLLOQUIUM SERIES ON GENOMIC AND MOLECULAR MEDICINE #1

ABSTRACT

Stroke is an important clinical problem because of its large contribution to mortality. The main causal and treatable risk factors for stroke include hypertension, diabetes mellitus, dyslipidemia, and smoking. In addition to these risk factors, recent studies have shown the importance of genetic factors and interactions between multiple genes and environmental factors. Genetic linkage analyses of families and sib-pairs as well as candidate gene association studies have implicated several loci and many candidate genes in predisposition to ischemic stroke, intracerebral hemorrhage, or subarachnoid hemorrhage. Recent genome-wide association studies identified various loci and genes that confer susceptibility to ischemic stroke or intracranial aneurysm. Such studies may provide insight into the function of implicated genes as well as into the role of genetic factors in the development of ischemic stroke, intracerebral hemorrhage, or subarachnoid hemorrhage.

KEYWORDS

ischemic stroke, atherothrombotic cerebral infarction, cardioembolic stroke, lacunar infarction, intracerebral hemorrhage, subarachnoid hemorrhage, genetics, polymorphism, linkage analysis, genome-wide association study

Contents

CHAPTER 1

Introduction

Stroke is a complex multifactorial disorder that is thought to result from an interaction between a person's genetic background and various environmental factors. It is a common and serious condition, with about 795,000 individuals experiencing a new or recurrent stroke and nearly 150,000 deaths from stroke-related causes in 2008 in the United States. The prevalence of stroke in the United States is 7 million. Of all such events, 87% are ischemic stroke, 10% are intracerebral hemorrhage, and 3% are subarachnoid hemorrhage (Roger et al., 2012) (Figure 1). Despite recent advances in acute stroke therapy, stroke remains the leading cause of severe disability and the third leading cause of death, after heart disease and cancer, in Western countries (Warlow et al., 2003). The identification of biomarkers of stroke risk is important both for risk prediction and for intervention to avert future events.

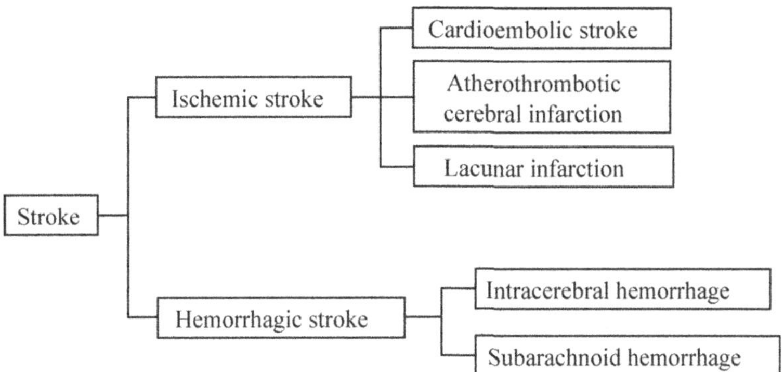

FIGURE 1: Classification of stroke. There are two types of stroke; ischemic stroke and hemorrhagic stroke. Ischemic stroke includes atherothrombotic cerebral infarction, cardioembolic stroke, and lacunar infarction. Hemorrhagic stroke includes intracerebral hemorrhage and subarachnoid hemorrhage.

• • • •

CHAPTER 2

Genetics of Stroke

Ischemic and hemorrhagic stroke may have both shared and different determinants, although the genetic variants that influence these clinical conditions are probably different. Studies with twins, siblings, and families have provided substantial evidence for heritability of common forms of stroke (Bak et al., 2002), but the genetic determinants remain largely unknown. Specific mutations in several monogenic stroke disorders have been identified (Meschia and Worrall, 2003). Although these observations provide insight into the pathophysiological processes of stroke, these mutations are rare and do not contribute substantially to stroke risk in the general population.

A family history of stroke is regarded as an important risk factor for this disease (Williams et al., 2001). A positive family history might be the result of shared genes, a shared environment, or both. Despite the identification of rare Mendelian stroke syndromes in humans (Tournier-Lasserve et al., 1991; Hassan and Markus, 2000), many candidate gene association studies for common forms of stroke have produced few consistent results and data on the genetic epidemiology of stroke are conflicting (Hassan and Markus, 2000; Hademenos et al., 2001). The incidence of ischemic stroke, intracerebral hemorrhage, or subarachnoid hemorrhage differs among ethnic groups, which may be attributable to differences in the distribution and frequency of genetic polymorphisms as well as in environmental factors such as diet, exercise, and other lifestyle aspects. Given that some gene polymorphisms characteristic of specific ethnic groups may be related to stroke, it is necessary to examine the relations of gene polymorphisms to stroke in each ethnic group.

• • • •

CHAPTER 3

Single-Gene Disorders Associated with Stroke

Several conditions in which stroke occurs are inherited in a classical Mendelian pattern as autosomal dominant, autosomal recessive, or X-linked disorders (Natowicz and Kelley, 1987; Hassan and Markus, 2000; Tournier-Lasserve, 2002). In most of these conditions, stroke is just one component of the disease phenotype, but in others, it is the prominent or sole clinical manifestation (Tournier-Lasserve, 2002). Studies of some Mendelian forms of stroke have identified the genes responsible (Carr et al., 2002; Markus, 2011).

Cerebral arteriopathy, autosomal dominant, with subcortical infarcts and leukoencephalopathy (CADASIL), an autosomal dominant form of stroke, has been well characterized genetically and shown to be attributable to mutation of the notch 3 gene (*NOTCH3*). Linkage analysis thus mapped the responsible gene to a defined region of human chromosome 19p13 (Tournier-Lasserve et al., 1993; Chabriat et al., 1995), the gene was then isolated by positional cloning, and the mutation was identified and its functional role confirmed (Joutel et al., 1996; Joutel et al., 1997). A striking feature of CADASIL is that disease severity is highly variable even within families. Recent data suggest the importance of both genetic and environmental modifying factors. Confluent white matter hyperintensities (WMH) on magnetic resonance imaging (MRI) are a key feature in CADASIL. A family study measuring WMH volume within a CADASIL population demonstrated a heritability of 63%, suggesting that other genes interact with *NOTCH3* to modulate the disease phenotype (Opherk et al., 2006).

Genes causing rare monogenic forms of small-vessel disease have recently been identified. Cerebral autosomal recessive arteriopathy with subcortical infarcts and leucoencephalopathy (CARASIL) causes lacunar stroke and early onset vascular dementia. Individuals also typically have alopecia beginning in the second decade, and spondylosis in the second or third decade. It has been shown to result from mutations in the HtrA serine peptidase 1 gene (*HTRA1*) which is involved in transforming growth factor, beta 1 signalling (Hara et al., 2009).

Two deadly forms of inherited intracerebral hemorrhage have been described in the Dutch and Icelandic populations (Hademenos et al., 2001). Cerebral hemorrhage with amyloidosis, hereditary, Dutch type (HCHWA-D) is due to a mutation in the amyloid beta precursor protein gene (*APP*) (Levy et al., 1990). The Icelandic form of this condition (HCHWA-I) is due to mutations in the gene coding for cystatin C (*CST3*), a serine protease inhibitor (Jensson et al., 1987; Palsdottir et al., 1988). These disorders are characterized by the development of cerebral hemorrhage at an age of 40 to 50 years for HCHWA-D and 20 to 30 years for HCHWA-I. Both are associated with amyloid deposition in cortical and leptomeningeal arterioles (Hademenos et al., 2001). Mutations of the integral membrane protein 2B gene (*ITM2B*) have also been shown to result in autosomal dominant amyloid angiopathies, which lead to cerebral hemorrhage, vascular dementia, or both (Vidal et al., 1999). The KRIT1, ankyrin repeat containing gene (*KRIT1*), has also been identified as one of the genes responsible for cavernous angiomas (Laberge-le Couteulx et al., 1999).

Autosomal dominant retinal vasculopathy with cerebral leucodystrophy is a microvascular endotheliopathy presenting with visual loss, stroke, and dementia, with onset in the middle age. C-terminal frame-shift mutations in the three prime repair exonuclease 1 gene (*TREX1*), which is ubiquitously expressed in the mammalian cells, were identified (Richards et al., 2007). These truncated proteins retain exonuclease activity but lose normal perinuclear localization (Markus, 2011).

Mutations in the collagen, type IV, alpha 1 gene (*COL4A1*) affecting glycine residues that are in close proximity in exons 24 and 25 within the triple helix domain of the protein cause a syndrome of hereditary angiopathy with nephropathy, aneurysm, and muscle cramps (HANAC). Patients with HANAC have disruption of the basement membrane systemically. The nephropathy causes hematuria and renal cysts. The muscle cramps occurs with or without elevations in serum creatine kinase. Patients have a characteristic retinal arteriopathy. There has been further characterization of the neurovascular phenotype in a series of families (Alamowitch et al., 2009). Infarctions not related to cardiac or large vessel pathology occur at an early age. Patients may be predisposed to posttraumatic hemorrhage. Intracranial aneurysms are characteristically localized to various levels of the carotid siphon (Meschia, 2011).

Another Mendelian condition associated with stroke is mitochondrial myopathy, encephalopathy, lactic acidosis, and strokelike episodes (MELAS), a genetically heterogeneous mitochondrial disorder with a variable clinical phenotype. It is accompanied by features of central nervous system involvement, including seizures, hemiparesis, hemianopsia, cortical blindness, and episodic vomiting (Pavlakis et al., 1984; Montagna et al., 1988). This syndrome has been attributed to single nucleotide mutations in mitochondrial DNA. The mutations are usually, but not exclusively, missense and lie within the $tRNA^{Leu(UUR)}$ gene, with an A→G transition at position 3243 (Enter et al., 1991) and a T→C transition at position 3271 (Sakuta et al., 1993) being most frequently reported. Individuals who have inherited one (Ciafaloni et al., 1992; Macmillan et al., 1993) or both (Pulkes et al., 2000) of these mutations have a greater predisposition to stroke (Carr et al., 2002).

Ischemic stroke is occasionally attributable to an underlying connective tissue disorder that results in arterial dissection (Carr et al., 2002). In Marfan syndrome, extension of aortic dissection into the common carotid artery can occur and result in stroke (Spittell et al. 1993). Defects in collagen synthesis in Ehlers–Danlos syndrome type IV can predispose affected individuals to spontaneous dissection of the extracranial carotid and vertebral arteries (Schievink et al., 1990). Fabry disease is an X-linked disorder caused by a deficiency of galactosidase, alpha and is associated with a high risk of both stroke and coronary heart disease (Crutchfield et al., 1998).

Identification of the genes responsible for Mendelian forms of stroke by reverse genetics has provided new insights into the pathophysiology of stroke (Tournier-Lasserve, 2002). These observations constitute the basis for clinically useful molecular diagnostic tests. Despite their low prevalence, monogenic conditions should always be considered in young patients who present with stroke or in patients of any age with no evidence of vascular risk factors, especially when there is a family history. Indeed, the risk of stroke both in individuals known to have the mutated gene and in their relatives is high. For example, in the case of an autosomal dominant disorder with complete penetrance, all persons who carry the mutated gene will have a stroke, as will half of their first-degree relatives (Tournier-Lasserve, 2002).

• • • •

CHAPTER 4

Genetics of Common Forms of Stroke

The etiology of common forms of stroke is multifactorial and includes both genetic and environmental factors. Studies with families have estimated that the relative risk of stroke in a first-degree relative of an individual who has a stroke is between 1.5 and 2.5. Such a risk is low at the individual level and may not have practical clinical implications. However, this slight increase in the risk of stroke is important at the population level, because of the high incidence of stroke (Tournier-Lasserve, 2002). Identification of genetic variants that contribute to the increased risk of stroke is therefore clinically important.

Common forms of stroke are heterogeneous and most likely result in part from the additive or multiplicative effects of a wide spectrum of pathogenic alleles, each of which confers a small degree of risk. Some of these alleles may predispose individuals to specific types or subtypes of stroke by affecting certain intermediate factors that either lead to stroke, such as the intimal–medial thickness of the carotid artery, or have a direct independent effect on the risk of stroke. In addition, gene variants may also modulate the severity of stroke (Tournier-Lasserve, 2002).

In spite of the large number of studies that have identified genes or polymorphisms associated with stroke, only a small number of the findings of these studies have been confirmed by independent replication or in other ethnic groups. One reason for this inconsistency is that many studies have combined ischemic and hemorrhagic stroke and it is unlikely that these different pathological conditions are under the same genetic influences (Floßmann et al., 2004). Another reason is that, although ischemic stroke is a highly complex trait, few studies have assessed subtypes of ischemic stroke or have had sufficient statistical power to do so (Floßmann et al., 2004). Many studies have thus analyzed atherothrombotic cerebral infarction and cardioembolic stroke collectively as ischemic stroke; the former results from the development of atherosclerotic stenosis in carotid or vertebral arteries, whereas the latter is attributable to the obstruction of cerebral arteries by thrombi that are generated in the cardiac atrium or ventricle as a result of arrhythmia such as atrial fibrillation or of valvular or ischemic heart disease. However, it can be argued that atherosclerosis is also responsible for most cardioembolic strokes, given that many such events are the consequence of thrombus formation on the damaged endocardial surface in acute myocardial infarction or within a ventricular aneurysm caused by damage to cardiac muscle secondary to previous myocardial infarction (Gulcher et al., 2005). Most cardioembolic strokes used to be caused by atrial fibrillation secondary to mitral stenosis associated with rheumatic heart disease, but the incidence of childhood rheumatic fever has decreased markedly in the era of penicillin (Wilhelmsen et al., 2001a; Wilhelmsen et al., 2001b). Most cases of atrial fibrillation are now caused by cardiac damage secondary to coronary heart disease. A substantial proportion of cardioembolic strokes is therefore related to atherosclerosis of coronary arteries (Gulcher et al., 2005). The etiology of intracardiac thrombi is diverse, however, including lone atrial fibrillation and other arrhythmias, valvular heart disease, cardiomyopathies, as well as coronary heart disease. Atherothrombotic cerebral infarction and cardioembolic stroke are thus different disorders. Given that the effects of gene polymorphisms or haplotypes on the development of common forms of stroke are likely to be small, it is necessary to examine these disorders separately in order to identify associated genetic variants.

4.1 STRATEGIES FOR GENETIC ANALYSIS OF STROKE

There are two basic strategies for identifying genes that influence the predisposition to stroke: linkage analyses and association studies (Figure 2). Linkage analysis involves the proposition of a model to account for the pattern of inheritance of a phenotype observed in a pedigree. It determines whether the phenotypic locus is transmitted together with genetic markers of known chromosomal position. Association studies determine whether a certain allele occurs at a frequency higher than that expected by chance in individuals with a particular phenotype. Such an association is thus suggested by a statistically significant difference in the prevalence of alleles with respect to the phenotype. Association studies consisted of two strategies: the candidate gene approach and the genome-wide approach. The candidate gene approach involves the direct examination of whether an individual gene or genes might contribute to the trait of interest. This strategy has been widely applied to analysis of the possible association between genetic variants and disease outcome, with genes selected on the basis of a priori hypotheses regarding their potential etiologic role. It is characterized as a hypothesis-testing approach because of the biological observation supporting the proposed candidate gene. The candidate gene approach is not able, however, to identify disease-associated polymorphisms in unknown genes. In the genome-wide scan, single nucleotide polymorphisms (SNPs) or copy number variations (CNVs) distributed throughout the entire genome are used to identify genomic regions that harbor genes that influence the trait of interest with a detectable effect size. This is a hypothesis-generating approach, allowing the detection of previously unknown potential trait loci.

The recent development of high-density genotyping arrays has improved the resolution of unbiased genome-wide scans for common variants associated with multifactorial diseases. Currently,

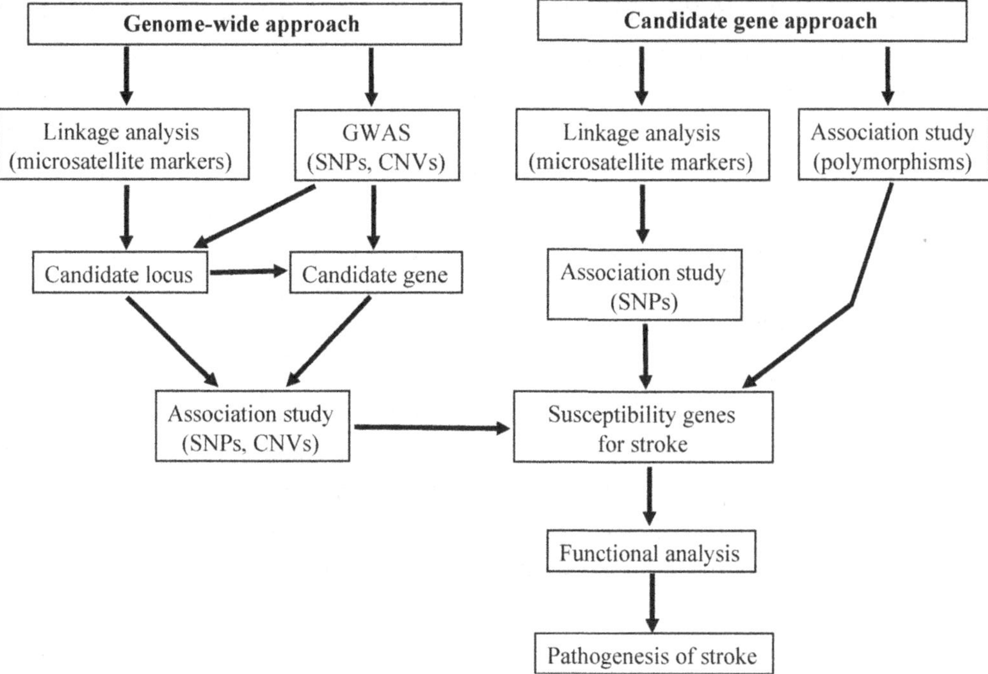

FIGURE 2: Strategies for identifying susceptibility genes for stroke. There are two basic strategies for identifying genes that influence common diseases or other complex traits, the genome-wide approach and the candidate gene approach, both of which rely on linkage analyses and association studies. In the genome-wide association study (GWAS), single nucleotide polymorphisms (SNPs) or copy number variations (CNVs) distributed throughout the entire genome are used to identify genomic regions that harbor genes that influence the trait of interest with a detectable effect size. The candidate gene approach involves the direct examination of whether an individual gene or genes might contribute to the trait of interest.

the genome-wide association study (GWAS) makes use of high-throughput genotyping technologies that include up to 4.3 million markers for SNPs and CNVs to examine their relation to clinical conditions or measurable traits. Until 25 February 2012, a Catalog of Published Genome-Wide Association Studies (National Human Genome Research Institute, NIH; http://www.genome.gov/gwastudies/) includes 1183 publications and 5910 SNPs associated with various diseases or traits, many in genes not previously suspected of having a role in the condition studied, and some in genomic regions containing no known genes. GWASs represent a substantial advance in the search for genetic variants that confer susceptibility to multifactorial polygenic diseases. GWASs, however, had disadvantages that previously available marker sets were designed to identify common alleles and were not well suited to study the effects of rare variants within a gene of interest.

4.2 MOLECULAR PATHOPHYSIOLOGY OF ISCHEMIC STROKE

Stroke is divided into two major varieties, ischemic and hemorrhagic stroke, with most (~87%) cases being ischemic. Ischemic stroke, which includes atherothrombotic cerebral infarction, cardioembolic stroke, and lacunar infarction (Figure 3), is characterized by a sudden decrease in blood flow to one or more central nervous system territories (Gulcher et al., 2005) and is a heterogeneous disease caused by different pathogenic mechanisms that include both environmental and genetic factors.

The pathophysiology of ischemic stroke is complex and involves numerous processes, including energy failure, loss of cell ion homeostasis, acidosis, increased intracellular calcium levels, excitotoxicity, free radical-mediated toxicity, generation of arachidonic acid products, cytokine-mediated cytotoxicity, complement activation, disruption of the blood brain barrier, activation of glial cells, and infiltration of leukocytes (Woodruff et al., 2011) (Figures 4 and 5). Within a few minutes of a cerebral ischemia, the core of brain tissue exposed to the most dramatic reduction of blood flow, is mortally injured, and subsequently undergoes necrotic cell death. This necrotic core is surrounded by a zone of less severely affected tissue which is rendered functionally silent by reduced blood flow but remains metabolically active (Majno and Joris, 1995; Broughton et al., 2009). Necrosis is morphologically characterized by initial cellular and organelle swelling, subsequent disruption of nuclear, organelle, and plasma membranes, disintegration of nuclear structure, and cytoplasmic organelles with extrusion of cell contents into the extracellular space (Majno and Joris, 1995;

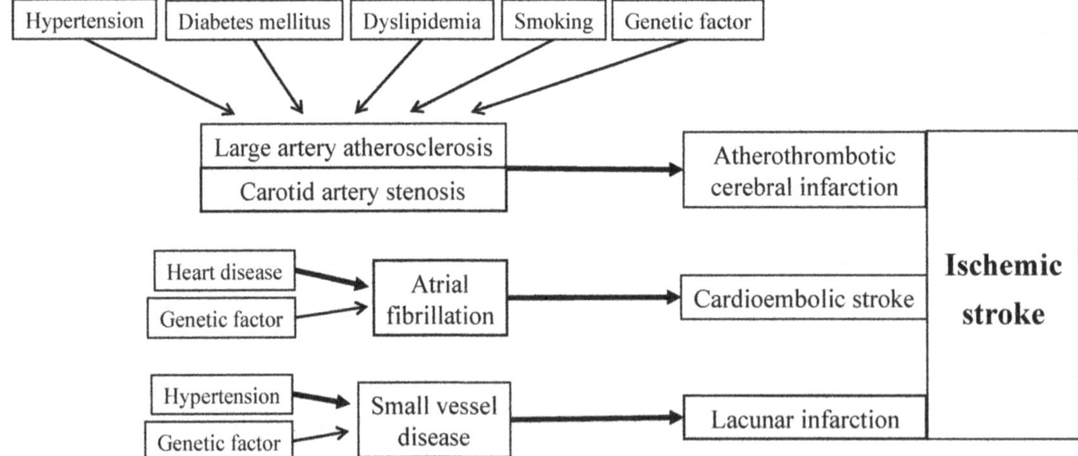

FIGURE 3: Etiology of ischemic stroke. Several risk factors have been shown for atherothrombotic cerebral infarction, cardioembolic stroke, or lacunar infarction, including hypertension, diabetes mellitus, and smoking. In addition to these conventional risk factors, genetic factors and interactions between multiple genes and environmental factors are important in the development of ischemic stroke.

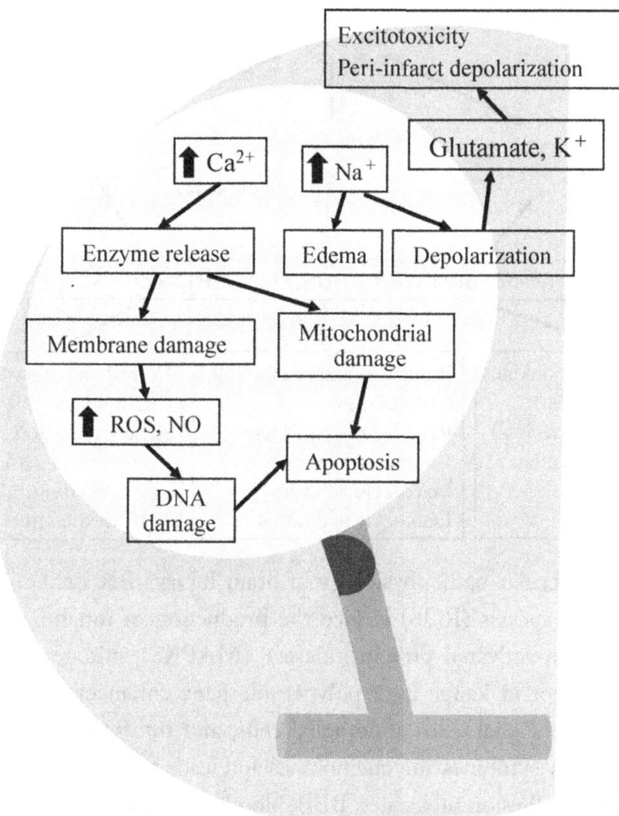

FIGURE 4: Proposed mechanisms of brain injury after ischemic stroke (Woodruff et al., 2011). Ischemia-induced energy failure leads to the depolarization of neurons. Activation of specific glutamate receptors dramatically increases intracellular Ca^{2+} and Na^{+}, and K^{+} is released into the extracellular space. Brain edema results from water shifts to the intracellular space. Increased levels of intracellular Ca^{2+} activate proteases, lipases, and endonucleases. Reactive oxygen species (ROS) are generated and damage membranes, mitochondria, and DNA, resulting in triggering cell death.

Broughton et al., 2009). The region bordering the infarction core, known as the ischemic penumbra, comprises as much as half of the total lesion volume during the initial stages of ischemia, and represents the region in which there is opportunity for salvage by post-stroke therapy (Ginsberg, 1997). Less severe ischemia, as occurs in the penumbra region of a focal infarction, evolves more slowly, and depends on the activation of specific genes and may ultimately result in apoptosis (Dirnagl et al., 1999; Lipton, 1999; Zheng and Yenari, 2004). Recent studies revealed that many neurons in the ischemic penumbra or peri-infarction zone may undergo apoptosis after several hours or days, and thus they are potentially recoverable for some time after the onset of stroke. In contrast to necrosis,

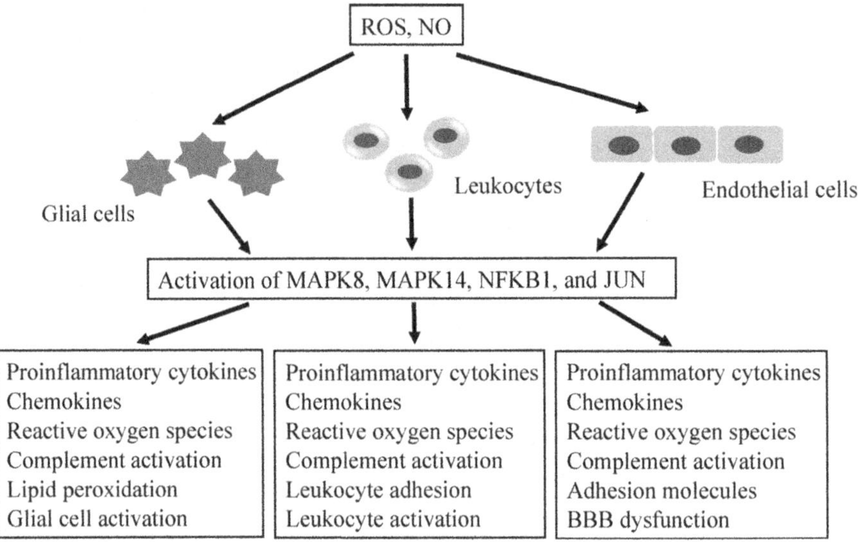

FIGURE 5: Proposed molecular pathophysiology of brain injury after ischemic stroke (Woodruff et al., 2011). Reactive oxygen species (ROS) induce the production of inflammatory mediators, resulting in activation of mitogen-activated protein kinase 8 (MAPK8), mitogen-activated protein kinase 14 (MAPK14), nuclear factor of kappa light polypeptide gene enhancer in B-cells 1 (NFKB1), and jun proto-oncogene (JUN) in glial cells, endothelial cells, and infiltrating leukocytes. This enhances secretion of proinflammatory cytokines and chemokines and leads to the invasion of leukocytes through upregulation of endothelial adhesion molecules. BBB, blood–brain barrier.

apoptosis appears to be an orderly process of energy-dependent programmed cell death to dispose of redundant cells. Cells undergoing apoptosis are dismantled in an organized way that minimizes damage and disruption to neighboring cells (Broughton et al., 2009). There are two general pathways for activation of apoptosis: the intrinsic and extrinsic pathways (Woodruff et al., 2011).

In rodent stroke models, neurons in the ischemic penumbra show morphological and molecular changes consistent with apoptosis, including caspase activation, expression of pro-apoptotic genes, and release of cytochrome c (Dirnagl et al., 1999). Signaling pathways involving hydrolysis of membrane phospholipids are implicated in neuronal apoptosis in stroke (Mattson, 2000). Cleavage of membrane sphingomyelin by acidic sphingomyelinase generates the lipid mediator ceramide. Focal cerebral ischemia in mice induces large increases in acidic sphingomyelinase activity and ceramide levels and the production of inflammatory cytokines (Yu et al., 2000). In mice lacking acidic sphingomyelinase, the production of cytokines is suppressed, brain damage is decreased, and symptoms are improved (Yu et al., 2000). Mice lacking phospholipase A2 also show decreased brain damage after focal cerebral ischemia, suggesting an important function of lipid mediators generated by this enzyme in ischemic neuronal injury (Bonventre et al., 1997).

4.3 MOLECULAR GENETICS OF ISCHEMIC STROKE

Studies with twins, siblings, and families have provided substantial evidence for stroke heritability (Bak et al., 2002). A mechanistic approach to the study of ischemic stroke, as advocated in the Trial of ORG 10172 in Acute Stroke Treatment (TOAST) study (Adams et al. 1993), is the best suited to genetic research (Morgan and Humphries 2005). This approach classifies ischemic stroke into five subtypes: (1) large-artery atherosclerosis; (2) small-vessel occlusion; (3) cardiogenic embolism; (4) stroke of other determined etiology; and (5) stroke of undetermined etiology. It was applied in a family history study of 1000 individuals with ischemic stroke and 800 controls (Jerrard-Dunne et al. 2003). This study found that a family history of vascular disease was a risk factor for both small-vessel occlusion and large-vessel atherosclerosis, but not for cardioembolic stroke or stroke of undetermined etiology. These findings suggest that genetic research may be most fruitful when focused on the former two subtypes of ischemic stroke (Morgan and Humphries, 2005).

Accurate phenotyping and performance of separate analyses according to stroke subtypes are thus essential. Focusing on particular stroke subtypes will likely make a study more efficient and markedly reduce the necessary sample sizes (Jerrard-Dunne et al. 2003). Another way of increasing the statistical power of a study may be to focus on early-onset cases, as is in any genetic predisposition. Family history studies, such as prospective twin studies (Brass et al., 1992), suggest that the genetic component of stroke is stronger in such individuals.

The main cause of ischemic stroke is atherothrombosis, with the principal and treatable risk factors including hypertension, diabetes mellitus, and dyslipidemia (Goldstein et al., 2001). In addition to these conventional risk factors, genetic variants are important in the pathogenesis of ischemic stroke (Hassan and Markus, 2000; Humphries and Morgan, 2004). Prediction of the risk for ischemic stroke beyond the usual clinical risk factors on the basis of genetic variants would be useful for deciding how aggressively to target the risk factors that are currently amenable to treatment. Furthermore, it might prompt earlier carotid imaging of patients at risk in order to detect asymptomatic carotid stenosis (Humphries and Morgan, 2004).

A whole-genome linkage analysis of families or sibling pairs showed that chromosomal region 5q12 was linked to ischemic stroke (Gretarsdottir et al., 2002). A large number of candidate gene association studies of unrelated individuals has identified many genes that are related to the prevalence of ischemic stroke (Table 1). Candidate gene association studies, however, have substantial limitations for detecting the genetic basis of stroke because this approach relies on selection of the genes for association studies based on either a biological hypothesis or the location of a particular gene in implicated linkage regions. In addition, most candidate gene association studies for stroke

TABLE 1: Genes shown to be related to ischemic stroke by linkage analyses or candidate gene association studies.

CHROMOSOMAL LOCUS	GENE NAME	GENE SYMBOL	REFERENCE
1p36.3	Methylenetetrahydrofolate reductase	*MTHFR*	Morita et al. (1998)
1p36.2	Natriuretic peptide A	*NPPA*	Rubattu et al. (2004)
1q21–q23	C-reactive protein, pentraxin-related	*CRP*	Morita et al. (2006)
1q23–q25	Selectin P	*SELP*	Zee et al. (2004)
1q25.2–q25.3	Prostaglandin-endoperoxide synthase 2	*PTGS2*	Cipollone et al. (2004)
2q14	Interleukin 1, beta	*IL1B*	Iacoviello et al. (2005)
2q14.2	Interleukin 1 receptor antagonist	*IL1RN*	Worrall et al. (2007)
3pter–p21	Chemokine (C-X3-C motif) receptor 1	*CX3CR1*	Lavergne et al. (2005)
3p25	Peroxisome proliferator-activated receptor gamma	*PPARG*	Lee et al. (2006)
4p16.3	Adducin 1	*ADD1*	Morrison et al. (2001)
4q28	Fibrinogen beta chain	*FGB*	Kessler et al. (1997)
4q28–q31	Fatty acid binding protein 2	*FABP2*	Carlsson et al. (2000)
5q12	Phosphodiesterase 4D, cAMP-specific	*PDE4D*	Gretarsdottir et al. (2003)
5q23–q31	Integrin, alpha 2	*ITGA2*	Carlsson et al. (1999)
5q31.1	Interleukin 4	*IL4*	Zee et al. (2004)
5q32–q33.1	Glutathione peroxidase 3	*GPX3*	Voetsch et al. (2007)

TABLE 1: (*continued*)

CHROMOSOMAL LOCUS	GENE NAME	GENE SYMBOL	REFERENCE
5q33–qter	Coagulation factor XII	*F12*	Santamaria et al. (2004)
6p25–p24	Coagulation factor XIII, A1 polypeptide	*F13A1*	Elbaz et al. (2000a)
6p21.3	Lymphotoxin alpha	*LTA*	Szolnoki et al. (2005)
6q22	c-Ros oncogene 1, receptor tyrosine kinase	*ROS1*	Yamada et al. (2008)
6q25.1	Estrogen receptor 1	*ESR1*	Shearman et al. (2005)
6q27	Lipoprotein, Lp(a)	*LPA*	Sun et al. (2003)
7p21	Interleukin 6	*IL6*	Pola et al. (2003) Yamada et al. (2006)
7q21.3	Paraoxonase 1	*PON1*	Voetsch et al. (2002)
7q21.3–q22	Serpin peptidase inhibitor, clade E, member 1	*SERPINE1*	Wiklund et al. (2005)
7q36	Nitric oxide synthase 3	*NOS3*	Elbaz et al. (2000b)
8p22	Lipoprotein lipase	*LPL*	Shimo-Nakanishi et al. (2001)
8p21–p12	Epoxide hydrolase 2, cytoplasmic	*EPHX2*	Fornage et al. (2005)
8p12	Plasminogen activator, tissue	*PLAT*	Saito et al. (2006)
9p21.3	CDKN2B antisense RNA 1	*CDKN2B-AS1*	Anderson et al. (2010)

TABLE 1: (*continued*)

CHROMOSOMAL LOCUS	GENE NAME	GENE SYMBOL	REFERENCE
9q31.1	ATP-binding cassette, sub-family A, member 1	*ABCA1*	Yamada et al. (2008)
11p11	Coagulation factor II	*F2*	De Stefano et al. (1998)
11q23	Apolipoprotein A-V	*APOA5*	Havasi et al. (2006)
12p13	Guanine nucleotide binding protein, beta polypeptide 3	*GNB3*	Morrison et al. (2001)
12p13	Sodium channel, nonvoltage-gated 1, alpha	*SCNN1A*	Hsieh et al. (2005)
13q12	Arachidonate 5-lipoxygenase-activating protein	*ALOX5AP*	Helgadottir et al. (2004)
14q11.2	Cathepsin G	*CTSG*	Herrmann et al. (2001)
14q22	Prostaglandin E receptor 2, 53kDa	*PTGER2*	Hegener et al. (2006)
16p11.2	Vitamin K epoxide reductase complex, subunit 1	*VKORC1*	Wang et al. (2006)
16q24	Cytochrome b-245, alpha polypeptide	*CYBA*	Ito et al. (2000)
17pter-p12	Glycoprotein Ib, alpha polypeptide	*GP1BA*	Baker et al. (2001)
17q21.32	Integrin, beta 3	*ITGB3*	Ridker et al. (1997)
17q23	Angiotensin I converting enzyme	*ACE*	Margaglione et al. (1996)
19p13.3	Thromboxane A2 receptor	*TBXA2R*	Kaneko et al. (2006)

TABLE 1: (*continued*)

CHROMOSOMAL LOCUS	GENE NAME	GENE SYMBOL	REFERENCE
19p13.3–p13.2	Intercellular adhesion molecule 1	*ICAM1*	Pola et al. (2003)
19p13.2	Low density lipoprotein receptor	*LDLR*	Frikke-Schmidt et al. (2004)
19q13.1	Transforming growth factor, beta 1	*TGFB1*	Kim and Lee (2006)
19q13.2	Apolipoprotein E	*APOE*	Kessler et al. (1997)
Xq28	Interleukin-1 receptor-associated kinase 1	*IRAK1*	Yamada et al. (2008)

have generated inconsistent or inconclusive results. Recent GWASs identified several loci and genes that confer susceptibility to ischemic stroke. The published results of GWASs for ischemic stroke are summarized in Table 2. In the following section, candidate genes for ischemic stroke of particular interest (*PDE4E, ALOX5AP, CDKN2B-AS1, NINJ2, PITX2, and ZFHX3*) are reviewed.

4.3.1 Phosphodiesterase 4D, cAMP-Specific Gene (*PDE4D*)

A genome-wide linkage study indicated that a gene on 5q12 may contribute to the risk of stroke (Gretarsdottir et al., 2002). A case-control study was performed to determine which linkage disequilibrium block within the linkage peak showed the strongest association with ischemic stroke. Markers in the alternative promoter region corresponding to one of the eight isoforms of PDE4D showed the strongest association (Gretarsdottir et al., 2003). The subtypes of ischemic stroke with the highest risk ratios were large-vessel occlusive disease and cardioembolic stroke; there was no association with small-vessel occlusive disease. The frequency of the most significant haplotype in each of these two patient subgroups was ~30%, and the relative risk was 1.98 for the two subgroups combined. A mutually exclusive haplotype that conferred protection was present in 28% of control individuals with a relative risk of 0.68. *PDE4D* variants thus conferred substantial risk for two forms of ischemic stroke that are related to atherosclerosis (Gulcher et al., 2005). Neither the risk nor protective haplotypes were associated with underlying missense or nonsense mutations, but they did correlate with the expression of *PDE4D* (Gretarsdottir et al., 2003).

TABLE 2: Chromosomal loci and genes shown to be related to ischemic stroke by genome-wide association studies.

CHROMOSOMAL LOCUS	dbSNP	NUCLEOTIDE SUBSTITUTION	GENE (NEARBY GENE)	PHENOTYPE	REFERENCE
4q25	rs2200733	C→T	*PITX2*	cardioembolic stroke	Gretarsdottir et al. (2008)
4q25	rs1906599	T→C	*PITX2*	cardioembolic stroke	ISGC & WTCCC2 (2012)
7p21.1	rs11984041	C→T	*HDAC9*	cerebral infarction	ISGC & WTCCC2 (2012)
11q12	rs9943582	C→T	*APLNR*	cerebral infarction	Hata et al. (2007)
12p13	rs12425791	G→A	*NINJ2*	cerebral infarction	Ikram et al. (2009)
12p13	rs11833579	G→A	*NINJ2*	cerebral infarction	Ikram et al. (2009)
14q23.1	rs2230500	G→A (Val374Ile)	*PRKCH*	lacunar infarction	Kubo et al. (2007)
16q22	rs7193343	T→C	*ZFHX3*	cardioembolic stroke	Gudbjartsson et al. (2009)
16q22.3	rs12932445	T→C	*ZFHX3*	cardioembolic stroke	ISGC & WTCCC2 (2012)
20p12.1	rs2208454	G→T	*MACROD2*	cerebral infarction	Debette et al. (2010)
22q13.3	rs6007897	A→G (Thr2268Ala)	*CELSR1*	cerebral infarction	Yamada et al. (2009)
22q13.3	rs4044210	A→G (Ile2107Val)	*CELSR1*	cerebral infarction	Yamada et al. (2009)

PDE4D degrades the second messenger cAMP (Fukumoto et al., 1999), which is a key signaling molecule in cell types that are important in the pathogenesis of atherosclerosis (Gulcher et al., 2005). A decrease in cAMP levels in vascular smooth muscle cells in vitro promoted the proliferation and migration of these cells, processes that are characteristic of atherosclerosis (Pan et al., 1994; Palmer et al., 1998; Fukumoto et al., 1999; Houslay and Adams, 2003). Inhibitors of PDE4 were found to block smooth muscle proliferation induced in the rat carotid artery by balloon injury (Indolfi et al., 1997; Indolfi et al., 2000). *PDE4D* is also expressed in activated macrophages and may therefore play a role in inflammation within atherosclerotic plaques, possibly contributing to atherogenesis or plaque instability, or both (Lusis, 2000; Libby, 2002; Naghavi et al., 2003). Increased activity of one or more isoforms of PDE4D resulting from dysregulation of transcript splicing or translation may thus increase the risk for ischemic stroke, with the decreased risk conferred by the identified protective haplotype possibly being due to a reduced activity of PDE4D (Gulcher et al., 2005).

Studies that have attempted to replicate this association of *PDE4D* with stroke have yielded diverse results (Markus and Alberts, 2006). In a U.K. population, no overall association was found with ischemic stroke, but possible associations were identified with cardioembolic stroke and large-artery stroke (Bevan et al. 2005). A U.S. study reported an association of *PDE4D* with ischemic stroke, especially with large-artery stroke (Meschia et al., 2005). In contrast, no association was found in a German stroke cohort (Lohmussaar et al. ,2005) or a Swedish stroke cohort of individuals aged < 75 years (Nilsson-Ardnor et al., 2005). A linkage study with a second Swedish population confirmed linkage of ischemic stroke to 5q12 (Nilsson-Ardnor et al., 2005), but no linkage was detected in an American population (Meschia et al., 2005). No association of *PDE4D* was found with carotid intimal–medial thickness (Bevan et al., 2005), suggesting that the gene does not exert its effects by accelerating early atherosclerosis (Markus and Alberts, 2006). A meta-analysis comprising 16 studies in 5216 cases and 6615 controls failed to detect the relation of *PDE4E* variants with ischemic stroke (Bevan et al., 2008). A multi-locus Bayesian meta-analysis including 14 data sets from populations of European descent and genotypes of 33 SNPs in 12,929 subjects (5994 cases and 6935 controls) confirmed no association despite the increase in statistical power (Newcombe et al., 2009). Although the regulation of intracellular cAMP concentration by PDE4D in vascular smooth muscle cells or even in macrophages may be a key determinant of stroke risk, the role of *PDE4D* variants in the pathogenesis of ischemic stroke remains unclear.

4.3.2 Arachidonate 5-Lipoxygenase-Activating Protein Gene (*ALOX5AP*)

A linkage and association study in Iceland demonstrated that *ALOX5AP* confers risk for both myocardial infarction and ischemic stroke (Helgadottir et al., 2004). The locus associated with myocardial infarction was initially mapped to chromosome 13q12 through a genome-wide linkage scan conducted on 296 families with this condition (Helgadottir et al., 2004). An independent linkage study of Icelandic stroke patients without myocardial infarction identified the same locus (Helgadottir et al., 2004). The haplotype defined by microsatellite markers that showed the strongest association with myocardial infarction covered a region containing *ALOX5AP* (Gulcher et al., 2005). A haplotype that spans *ALOX5AP* and is defined by four SNPs (HapA) was subsequently shown to be associated with myocardial infarction, with a relative risk of 1.8. The same haplotype was then found to confer risk for stroke in the Icelandic population with a relative risk of 1.7 (Gulcher et al., 2005). HapA is relatively common and is carried by 27% of Icelandic patients with stroke. Another haplotype within *ALOX5AP* (HapB) showed a significant association with myocardial infarction in British cohorts, with a relative risk of 2.0 (Helgadottir et al., 2004). The association of both haplotypes with stroke in Scottish individuals was subsequently demonstrated (Helgadottir et al., 2005).

ALOX5AP participates in the initial steps of leukotriene synthesis. Arachidonic acid is thus converted to leukotriene A4 by the action of arachidonate 5-lipoxygenase and its activating protein, ALOX5AP. Inflammatory lipid mediators, including leukotrienes B4, C4, D4, and E4 (Dixon et al., 1990), are then produced from leukotriene A4 by the action of leukotriene A4 hydrolase and leukotriene C4 synthase. The amount of leukotriene B4 synthesized by ionomycin-stimulated neutrophils from individuals with myocardial infarction was greater than that produced by those from control individuals (Helgadottir et al., 2004), supporting the notion that increased activity of the leukotriene pathway plays a role in the pathogenesis of myocardial infarction (Gulcher et al., 2005). Moreover, the observed difference in the release of leukotriene B4 was largely accounted for by carriers of HapA, whose cells produced more leukotriene B4 than did those from noncarriers. Although leukotriene B4 production was not measured in cells from patients with stroke, a similar increase would be expected, given that the HapA variant of *ALOX5AP* shows similar associations with ischemic stroke and myocardial infarction. Elevated levels of leukotriene B4 might contribute to atherogenesis or plaque instability by promoting inflammation at atherosclerotic plaques (Gulcher et al., 2005).

A role for up-regulation of the leukotriene pathway in atherosclerosis is further supported by the observation that expression of enzymes of the arachidonate 5-lipoxygenase pathway is increased in human atheromas, with the number of arachidonate 5-lipoxygenase-positive cells (macrophages, dendritic cells, mast cells, and neutrophils) being markedly increased in advanced lesions (Spanbroek et al., 2003). Furthermore, the arachidonate 5-lipoxygenase gene (*ALOX5*) has been implicated in the development of atherosclerosis in mice by the finding that the loss of only one *ALOX5* allele confers protection against atherosclerosis in animals deficient in the low density lipoprotein receptor (Mehrabian et al., 2002). An increased activity of the leukotriene biosynthetic pathway associated with specific *ALOX5AP* variants might thus promote the processes of atherogenesis and subsequent plaque instability, increasing the chance of ischemic stroke on the background of atherosclerosis (Gulcher et al., 2005).

Several groups have attempted to replicate the association of ischemic stroke with *ALOX5AP* variants (Markus and Alberts, 2006). Whereas the deCODE group replicated the association in a Scottish stroke population (Helgadottir et al., 2005), a case-control study in Germany reported a weak association with an *ALOX5AP* polymorphism (Lohmussaar et al., 2005), and no association in a case-control study and no linkage to this chromosomal region in a sibling-pair study were found in American populations (Bevan et al., 2005). A meta-analysis of 5194 stroke cases and 4566 controls showed significant heterogeneity among studies and failed to detect overall significant association (Zintzaras et al., 2009). Although the leukotriene biosynthetic pathway may play an important role in leukocyte chemotaxis and inflammatory responses that are key processes in atherosclerosis, the role of *ALOX5AP* polymorphisms in the pathogenesis of stroke remains elucidated.

4.3.3 CDKN2B Antisense RNA 1 Gene (*CDKN2B-AS1*)

Chromosome 9p21.3 locus was first discovered by GWASs to be a risk factor for coronary heart disease or myocardial infarction (Helgadottir et al., 2007; McPherson et al., 2007; Samani et al., 2007; Wellcome Trust Case Control Consortium, 2007). This locus was then shown to be associated with ischemic stroke (Matarin et al., 2008; Gschwendtner et al., 2009). A meta-analysis of the eight studies (9632 cases and 30,716 controls) confirmed the association of rs10757278 SNP at 9p21.3 with ischemic stroke. Similar analysis of two studies (5255 cases and 22,640 controls) also demonstrated the relation of rs1537378 at 9p21.3 to ischemic stroke (Anderson et al., 2010). The locus was also associated with intracranial and aortic aneurysms (Helgadottir et al., 2008). Recently, 9p21.3 has also been associated with platelet reactivity (Musunuru et al., 2010). The increased platelet reactivity may explain the association with myocardial infarction and ischemic stroke, however, it is not clear how increased platelet reactivity might relate to the formation of aneurysm. The 9p21.3 locus includes the CDKN2B antisense RNA 1 gene (*CDKN2B-AS1*), which alters expression of several genes related to cellular proliferation (Jarinova et al., 2009). Recent study also indicates that 9p21 polymorphisms influence inflammatory signaling (Harismendy et al., 2011) and vascular cell proliferation (Visel et al., 2010). These studies suggest that *CDKN2B-AS1* at 9p21.3 may be a susceptibility locus for myocardial infarction and ischemic stroke, although the underlying molecular mechanisms have not been determined definitively. An improved explanation for the pleiotropic effects of the 9p21.3 locus on multiple vascular beds should emerge in the near future (Cole and Meschia, 2011).

4.3.4 Ninjurin 2 Gene (*NINJ2*)

A prospective GWAS showed an association of two SNPs on chromosome 12p13 with ischemic stroke (Ikram et al., 2009). The primary study population, consisting of 19,602 subjects including 1544 incident strokes from four prospective white cohorts, identified two SNPs (rs11833579 and rs12425791) in the region of 12p13. Results of the initial GWAS suggested that a minor allele of each SNP increased the hazard ratio for total stroke by ~1.3 and for ischemic stroke by ~1.4. The corresponding population-attributable risks were 11% to 13% for total stroke and 14% to 17% for ischemic stroke. The association between the two SNPs and ischemic stroke was further tested in two independent replication samples, an African-American community-based cohort and a Dutch case-control sample, and rs12425791 was significantly associated with ischemic stroke, especially atherothrombotic stroke in both samples. This SNP was located close to *NINJ2*, which encodes an adhesion molecule expressed in glia that shows increased expression after nerve injury (Ikram et al., 2009). A meta-analysis combining data for 8637 cases and 8733 controls of European ancestry and a population-based genome-wide cohort study of 278 ischemic stroke among 22,054 participants did not replicate the relation of either SNP with ischemic stroke (Rosand et al., 2010). In addition, Italian (Lotta et al., 2010) and Swedish (Olsson et al., 2011) case-control studies also failed to show any association between the two genetic variants on 12p13 and ischemic stroke (Cole and Meschia, 2011). The relation of polymorphisms at 12p13 locus to ischemic stroke and the underlying molecular mechanisms thus remain unclear.

4.3.5 Paired-Like Homeodomain 2 Gene (*PITX2*) and Zinc Finger Homeobox 3 Gene (*ZFHX3*)

Atrial fibrillation is an important risk factor for cardioembolic stroke, and there are overlapping genetic risk factors for these two conditions. A multistage study in 1661 cases with ischemic stroke and 10,815 controls in the discovery phase showed that two SNPs (rs2200733 and rs10033464) at 4q25 near *PITX2* were significantly associated with cardioembolic stroke (Gretarsdottir et al., 2008). These SNPs had already been associated with atrial fibrillation. Another association study involving 4199 cases of ischemic stroke and 3750 controls showed an association between the 4q25 locus and cardioembolic stroke (Lemmens et al., 2010).

Overlapping genetic risk for atrial fibrillation and ischemic stroke was further observed for a sequence variant in the zinc finger homeobox 3 gene (*ZFHX3*) on 16q22 (Gudbjartsson et al. 2009). The GWAS of samples from Iceland, Norway, and the United States observed that rs7193343 was significantly associated with atrial fibrillation. This variant was also associated with cardioembolic stroke. *ZFHX3* encodes a transcription factor with multiple homeodomains and zinc finger motifs and regulates myogenic and neuronal differentiation (Meschia, 2011).

4.4 MOLECULAR PATHOPHYSIOLOGY OF INTRACEREBRAL HEMORRHAGE

Intracerebral hemorrhage is responsible for ~10% of all strokes, including a large proportion of fatal or severe cases. Advancing age and hypertension are the most important risk factors for intracerebral hemorrhage. Intracerebral hemorrhage is usually attributed to hypertensive small-vessel disease, with the most common sites of hemorrhage being the basal ganglia, cerebellum, and pons (Figure 6).

Intraparenchymal bleeding results from the rupture of the small penetrating arteries that originate from basilar arteries or the anterior, middle, or posterior cerebral arteries. Degenerative changes in the vessel wall induced by chronic hypertension reduce compliance and increase the likelihood of spontaneous rupture. Electron-microscopical studies suggest that most bleeding occurs at

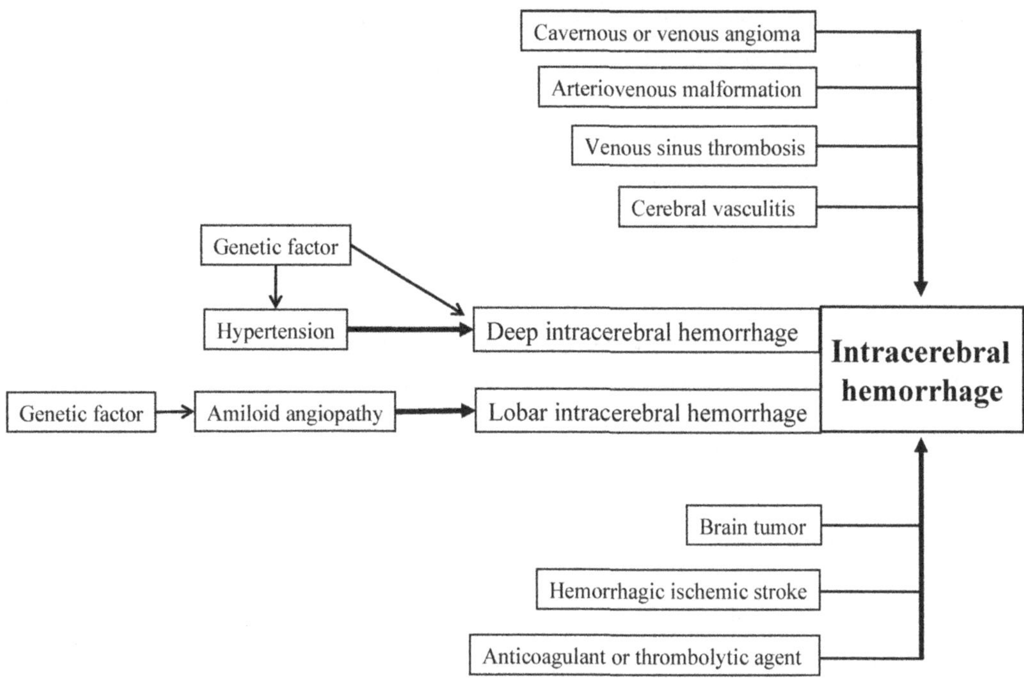

FIGURE 6: Etiology of intracerebral hemorrhage. Intracerebral hemorrhage includes hypertensive deep cerebral hemorrhage and lober cerebral hemorrhage caused by amyloid angiopathy. Intracerebral hemorrhage is also occurred by various conditions, including cavernous or venous angioma, arteriovenous malformation, venous sinus thrombosis, cerebral vasculitis, brain tumor, hemorrhagic ischemic stroke, and anticoagulant or thrombolytic agent. Although hypertension is the most important risk factor for intracerebral hemorrhagic, genetic factors are also involved in the development of this condition.

or near the bifurcation of affected arteries, where prominent degeneration of the media and smooth muscles can be observed (Cole and Yates, 1967; Qureshi et al., 2001).

During intracerebral hemorrhage, rapid accumulation of blood within brain parenchyma leads to disruption of normal anatomy and an increase in local pressure. Depending on the expansion of hematoma, the primary damage occurs within minutes to hours from the onset of bleeding and is primarily the result of mechanical damage associated with the mass effect (Qureshi et al., 2009). Secondary damage is attributable to the presence of intraparenchymal blood and may depend on the initial hematoma volume, age, or ventricular volume (Qureshi et al., 2009). Secondary damage may occur through many parallel pathological pathways, including cytotoxicity of blood (Wagner et al., 2003; Xi et al., 2006), hypermetabolism (Ardizzone et al., 2004), excitotoxicity (Qureshi et al., 2003), spreading depression (Mun-Bryce et al., 2001), and oxidative stress and inflammation (Hickenbottom et al., 1999; Gong et al., 2000; Xue and Del Bigio, 2000; Aronowski and Hall, 2005; Tang et al., 2005; Xi et al., 2006; Wang et al., 2007; Wang and Doré, 2007) (Figure 7). These pathological events lead to irreversible disruption of the components of the neurovascular unit, constituting gray and white matters, and is followed by disruption of blood–brain barrier and deadly brain edema with massive brain cell death (Felberg et al., 2002; Huang et al., 2002; Wang et al., 2002; Qureshi et al., 2003; Wagner et al., 2003; Aronowski and Hall, 2005; Xi et al., 2006). Whereas inflammatory mediators generated locally in response to brain injury augment damage caused by intracerebral hemorrhage (secondary injury), the involvement of inflammatory cells, such as microglia and macrophages, is vital for removal or cleanup of cellular debris from hematoma, the source of ongoing inflammation (Zhao et al., 2007a). The timely removal of damaged tissue is essential for reducing the length of deleterious pathological process and thereby allowing for faster and more efficient recovery (Aronowski and Zhao, 2011).

The inflammatory signaling in a hemorrhage-affected brain involves a transcription factor, nuclear factor of kappa light polypeptide gene enhancer in B-cells 1 (NFKB1) (Hickenbottom et al., 1999; Zhao et al., 2007b). The target genes of NFKB1 include those for various adhesion molecules including intercellular adhesion molecule 1; proinflammatory cytokines including interleukin 1, beta and tumor necrosis factor; chemokines; matrix metallopeptidases including matrix metallopeptidase 9; immune receptors; acute phase proteins; cell surface receptors; and inflammatory enzymes including nitric oxide synthase 2, inducible, prostaglandin-endoperoxide synthase 2, and phospholipase A2 (Aronowski and Zhao, 2011). Reactive oxygen species act as important signaling molecules in activation of NFKB1. This property of NFKB1 may, in part, explain how oxidative stress enhances inflammation after intracerebral hemorrhage (Aronowski and Zhao, 2011). Experimental studies demonstrate that NFKB1 is activated in a hemorrhage-affected hemisphere as early as 15 minutes after the onset of intracerebral hemorrhage, reaches maximum between one and three days, and remains elevated for weeks (Zhao et al., 2007b). Interleukin 1, beta, tumor necrosis

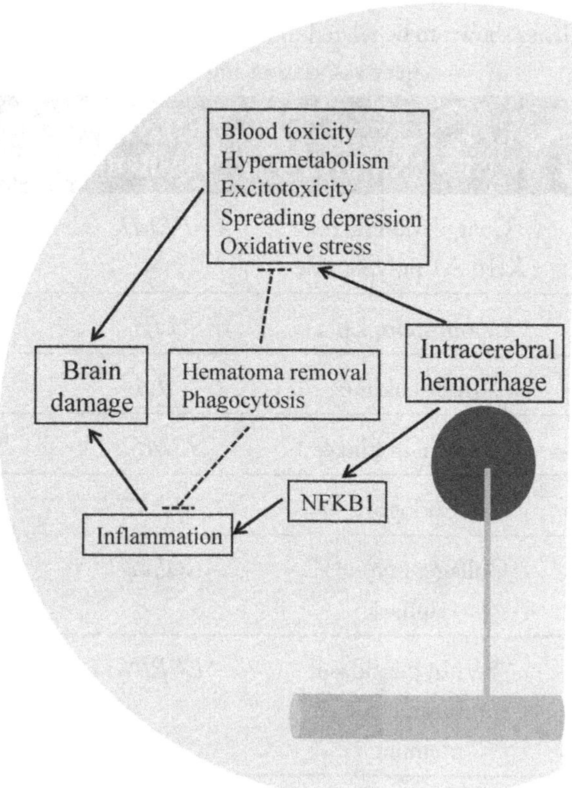

FIGURE 7: Proposed mechanisms of secondary brain injury after intracerebral hemorrhage (Aronowski and Zhao, 2011). Intracerebral hemorrhage activates nuclear factor of kappa light polypeptide gene enhancer in B-cells 1 (NFKB1), which then induces inflammation that leads to secondary brain damage. Intracerebral hemorrhage also induces blood toxicity, hypermetabolism, excitotoxicity, spreading depression, and oxidative stress, which result in brain damage. Hematoma removal and phagocytosis suppress these adverse events, leading to reduction of brain damage.

factor, and matrix metallopeptidase 9, which are upregulated by NFKB1, are shown to be involved in hemorrhage-mediated brain injury (Masada et al., 2001; Mayne et al., 2001; Rosenberg, 2002; Tang et al., 2004; Wang and Doré, 2007).

4.5 MOLECULAR GENETICS OF INTRACEREBRAL HEMORRHAGE

Familial aggregation of cases of intracerebral hemorrhage was demonstrated in a prospective study in North Carolina in the United States, which found that 10% of affected individuals had a family

TABLE 3: Genes shown to be related to intracerebral hemorrhage by candidate gene association studies.

CHROMOSOMAL LOCUS	GENE NAME	GENE SYMBOL	REFERENCE
6p25–p24	Coagulation factor XIII, A1 polypeptide	*F13A1*	Reiner et al. (2001)
6q27	Lipoprotein, Lp(a)	*LPA*	Sun et al. (2003)
7p21	Interleukin 6	*IL6*	Yamada et al. (2006)
7q11.23	Lim domain kinase 1	*LIMK1*	Yamada et al. (2008)
9q34.1	Endoglin	*ENG*	Alberts et al. (1997)
13q34	Collagen, type IV, alpha 1	*COL4A1*	Gould et al. (2006)
14q32.1	Serpin peptidase inhibitor, clade A, member 3	*SERPINA3*	Vila et al. (2000)
16p11.2	Vitamin K epoxide reductase complex, subunit 1	*VKORC1*	Wang et al. (2006)
17q23	Angiotensin I converting enzyme	*ACE*	Slowik et al. (2004b)
17q23-qter	Apolipoprotein H	*APOH*	Xia et al. (2004)
19q13.2	Apolipoprotein E	*APOE*	O'Donnell et al. (2000)

history of intracerebral hemorrhage (Alberts et al., 2002). No significant clinical demographic differences separated affected individuals with or without a family history of intracerebral hemorrhage. Genetic factors may influence not only the development of intracerebral hemorrhage but also the prevalence of certain risk factors for this condition, such as hypertension. Furthermore, such genetic factors may interact with environmental factors such as diet and cigarette smoking.

In some individuals with intracerebral hemorrhage, the hemorrhage is lobar in location, such as in the frontal, parietal, temporal, or occipital cortex, and such patients often do not have hypertension (Massaro et al., 1991). This category of hemorrhage, referred to as lobar intracerebral hemorrhage, may represent a distinct pathogenetic subgroup (Sacco, 2000) (Figure 6). The occurrence of lobar intracerebral hemorrhage was shown to be associated with the ε2 and ε4 alleles of the apolipoprotein E gene (*APOE*) (O'Donnell et al., 2000; Woo et al., 2005). These relations, particularly that with ε4, are presumably due to the association of *APOE* with cerebral amyloid angiopathy (Greenberg et al., 1996).

Various candidate gene association studies of unrelated individuals have identified genes that are related to intracerebral hemorrhage (Table 3). The etiology of intracerebral hemorrhage is complex and the genetic determinants of this condition are still largely unknown. In the following sections, candidate genes for intracerebral hemorrhage that are of particular interest (*APOE* and *COL4A1*) are reviewed.

4.5.1 Apolipoprotein E Gene (*APOE*)

Cerebral amyloid angiopathy is a frequent cause of lobar intracerebral hemorrhage (Sacco, 2000). The main pathological feature of this condition is the infiltration of cortical vessels by amyloid beta, a homogeneous eosinophilic substance found in the brain of elderly individuals and an important component of the senile plaques in patients with Alzheimer's disease. The incidence of lobar intracerebral hemorrhage due to cerebral amyloid angiopathy increases markedly with age, with most affected individuals being over the age of 60 years and many having antecedent memory loss. Patients with cerebral amyloid angiopathy and intracerebral hemorrhage have a lower mortality rate and a greater risk of recurrence than do those with other types of intracerebral hemorrhage (Sacco, 2000).

The ε4 and ε2 alleles of APOE were identified as predictors of recurrent lobar intracerebral hemorrhage in patients with cerebral amyloid angiopathy (O'Donnell et al., 2000). The risk of recurrence at two years was 28% for carriers of the ε2 or ε4 alleles, compared with 10% for patients with the ε3 allele. The presence of the ε2 or ε4 alleles is thus considered a potent risk factor for recurrence (Sacco, 2000).

Other studies have shown that the frequency of the ε4 allele of *APOE* is increased in patients with cerebral amyloid angiopathy, whereas the ε2 allele may be associated with an increased risk of intracerebral hemorrhage in individuals with this condition (Greenberg et al., 1995; Nicoll et al., 1997; Greenberg et al., 1998; O'Donnell et al., 2000). These observations may be partially explained by the association of the ε4 allele with Alzheimer's disease, which is present in up to 50% of patients with cerebral amyloid angiopathy. In addition, the age of onset of intracerebral hemorrhage in individuals with cerebral amyloid angiopathy was found to be earlier in carriers of the ε4 allele of *APOE* than in carriers of other alleles of this gene (Greenberg et al., 1996).

4.5.2 Collagen, Type IV, Alpha 1 Gene (*COL4A1*)

Mutations in *COL4A1* have been detected in families with cerebral small-vessel disease (Gould et al., 2005; Gould et al., 2006). *COL4A1* was initially identified as the causative gene in a mouse mutant with perinatal cerebral hemorrhage and porencephaly (Gould et al., 2005). Mice heterozygous for the responsible mutation develop recurrent hemorrhage in the basal ganglia, a typical site of intracerebral hemorrhage in hypertensive patients. Subsequent analysis of families with porencephaly and cerebral small-vessel disease revealed several mutations of human *COL4A1* in affected individuals (Gould et al., 2005; Gould et al., 2006; van der Knaap et al., 2006).

Type IV collagens are an integral component of the vascular basement membrane. COL4A1 and COL4A2, the most abundant type IV collagens, form heterotrimers. Repeated Gly-Pro-X motifs of these collagen molecules are required for formation of a triple helix during collagen assembly, and most mutations identified in *COL4A1* affect Gly residues within these motifs. It was therefore hypothesized that *COL4A1* mutations interfere with triple-helix formation or heterotrimer secretion. Indeed, analysis of heterozygous embryonic tissue has suggested that the mutations inhibit collagen secretion into the basement membrane. Ultrastructural abnormalities in capillaries of carriers of *COL4A1* mutations are indicative of disordered assembly of the basement membrane (Dichgans and Hegele, 2007).

The phenotypic spectrum associated with *COL4A1* mutations is broad and strongly related to small-vessel disease. Key features include leukoencephalopathy, microhemorrhages, and clinically overt hemorrhage. The structural compromise of small blood vessels is reflected by the observations that birth trauma, brain trauma, or oral anticoagulants may trigger intracerebral hemorrhage in carriers of *COL4A1* mutations (Gould et al., 2005; Gould et al., 2006). Genes that encode proteins associated with the vascular basement membrane are thus potential candidates for causative agents of intracerebral hemorrhage and leukoencephalopathy (Dichgans and Hegele, 2007).

4.6 MOLECULAR PATHOPHYSIOLOGY OF INTRACRANIAL ANEURYSM AND SUBARACHNOID HEMORRHAGE

Subarachnoid hemorrhage is most commonly caused by rupture of an aneurysm on an intracranial artery (Ruigrok et al., 2005). About 2% of the general population has an intracranial aneurysm (Rinkel et al., 1998). Rupture of an aneurysm is most common between 40 and 60 years of age and prognosis after rupture is poor, with half of affected individuals dying within one month and 20% remaining dependent on support for activities of daily life (Longstreth et al., 1993; Inagawa et al., 1995; Hop et al., 1997). The incidence of aneurysmal subarachnoid hemorrhage in the general

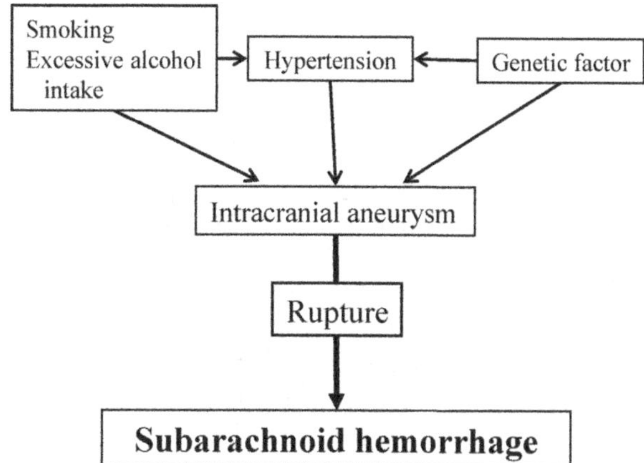

FIGURE 8: Etiology of subarachnoid hemorrhage. Hypertension, smoking, and excessive alcohol intake are risk factors for subarachnoid hemorrhage. In addition to these factors, genetic factors are important in the development of intracranial aneurysm and subarachnoid hemorrhage.

population is low (about 8 per 100,000 person-years) (Linn et al., 1996), but the young age at onset and the poor prognosis mean that the loss of productive life-years is similar to that associated with ischemic stroke (Johnston et al., 1998).

The pathogenesis of subarachnoid hemorrhage from a ruptured aneurysm has not been fully understood. Hemodynamic factors and structural properties of the arterial wall may contribute to the development of intracranial aneurysms, but the trigger factors remain unknown (Figure 8). Disruption of the extracellular matrix is likely to be a factor in the pathophysiology given that intracranial aneurysms are associated with heritable disorders of connective tissue and the extracellular matrix (Chapman et al. ,1992; Schievink et al., 1994; Ruigrok et al., 2005). Moreover, the amount of structural proteins of the extracellular matrix has been found to be reduced in the intracranial arterial wall of many ruptured intracranial aneurysms as well as in skin biopsies and intracranial and extracranial arteries of patients with aneurysms (Neil-Dwyer et al., 1983; Hegedus, 1984; Ostergaard and Oxlund, 1987; Ostergaard et al., 1987; Chyatte et al., 1990; Skirgaudas et al., 1996; van den Berg et al., 1997).

Recent studies suggested that intracranial aneurysm is a chronic inflammatory disease at the bifurcation site of cerebral arterial walls (Aoki and Nishimura, 2010) (Figure 9). Many inflammatory cells were observed in human samples from surgically dissected intracranial aneurysm walls (Chyatte et al., 1999). In the intracranial aneurysms of rodent models, macrophages were the main inflammatory cells, and the number of macrophages gradually increased during aneurysm formation, suggesting the involvement of macrophage-mediated inflammation in the formation of intracranial

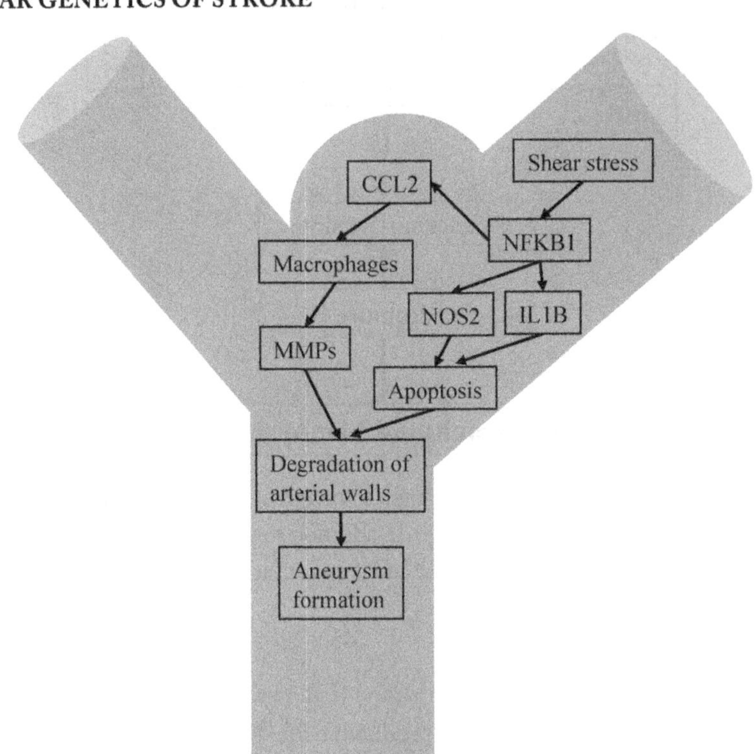

FIGURE 9: Proposed mechanisms of the formation of intracranial aneurysm (Aoki and Nishimura, 2010). Shear stress activates NFKB1 in endothelial cells. Activation of NFKB1 increases the expression of NOS2 and IL1B, resulting in acceleration of apoptosis of arterial wall cells. Activation of NFKB1 also increases release of CCL2 that recruits macropahages to the arterial walls. Macropahages produce matrix metallopeptidases (MMP), especially MMP9, resulting in degradation of arterial walls. Weakening of arterial walls by chronic inflammation may thus be a crucial mechanism of the formation of intracranial aneurysm. CCL2, chemokine (C–C motif) ligand 2; NFKB1, nuclear factor of kappa light polypeptide gene enhancer in B-cells 1; NOS2, nitric oxide synthase 2, inducible; IL1B, interleukin 1, beta.

aneurysm (Aoki et al., 2007). Chemokine (C–C motif) ligand 2 (CCL2) is the major chemoattractant for macrophages and contributes to recruitment of macrophages to the inflammatory site. The CCL2 mRNA was expressed in human intracranial aneurysm walls, mainly in endothelial cells, at the early stage of aneurysm formation, and CCL2 deficiency resulted in reduced incidence of intracranial aneurysm through suppression of macrophage infiltration to the arterial walls (Aoki et al., 2009). These observations suggest that macrophages contribute to the formation of intracranial aneurysm and that chronic inflammation is involved in the pathogenesis of intracranial aneurysm (Aoki and Nishimura, 2010).

Degeneration of the extracellular matrix is the major pathological feature of intracranial aneurysm. Degeneration of the extracellular matrix may be related to enlargement and rupture of the aneurysm as a result of weakening of arterial walls. Increased collagenolytic activity and reduced collagen content were detected in human intracranial aneurysm walls (Gaetani et al., 1997; Gaetani et al., 1998). Matrix metallopeptidases (MMP), especially MMP9, may have an important role in the formation of intracranial aneurysm (Chyatte and Lewis, 1997; Bruno et al., 1998; Aoki et al., 2007). The expression of MMP9 was increased in macrophages infiltrating aneurysm walls. Inhibition of the MMP9 activity by a selective inhibitor resulted in reduced incidence of intracranial aneurysm, suggesting the important role of MMP9 in the formation of intracranial aneurysm (Aoki et al., 2007). Apoptosis of medial smooth muscle cells is a possible mechanism of degeneration of the media (Kondo et al., 1998). Among molecules related to apoptosis cell signaling, nitric oxide synthase 2, inducible (NOS2) and interleukin 1, beta (IL1B) may contribute to the formation of intracranial aneurysm. NOS2 and IL1B were expressed in the media of aneurysm walls. Deficiency of NOS2 or IL1B resulted in reduced aneurysm formation by suppression of apoptosis in medial smooth muscle cells of the arterial walls, suggesting the involvement of apoptotic cell death in the pathogenesis of intracranial aneurysm (Fukuda et al., 2000; Sadamasa et al., 2003; Moriwaki et al., 2006; Guo et al., 2007).

Various proinflammatory genes, including those for MMPs, NOS2, and CCL2, have specific NFKB1 binding sites, suggesting the role of NFKB1 in the pathogenesis of aneurysm formation through the transcriptional regulation of proinflammatory genes (Ping et al., 1999; Lianxu et al., 2006). NFKB1 is activated by shear stress in endothelial cells, and regulates various proinflammatory genes as a shear stress-sensitive transcriptional factor (Khachigian et al., 1995; Ballermann et al., 1998; Davis et al., 2004; Orr et al., 2005). NFKB1 may thus link shear stress with chronic inflammation of aneurysm walls. Hypertension induced NFKB1 activation in the rat model of aortic aneurysm (Shiraya et al., 2006). These observations suggest that NFKB1 is a key mediator of the formation of intracranial aneurysm (Aoki and Nishimura, 2010).

Delayed cerebral ischemia is a common and serious complication following subarachnoid hemorrhage after ruptured intracranial aneurysm (de Rooij et al., 2007; Pluta et al., 2009) (Figure 10). Although this complication is at times reversible, it may develop into a cerebral infarction (Fergusen and Macdonald, 2007). Delayed cerebral ischemia occurs in 20% to 40% of patients and is associated with increased mortality and poor prognosis (Dorsch and King, 1994; Rabinstein et al., 2004; Macdonald et al., 2007). It is usually caused by a vasospasm (Vajkoczy et al., 2003), which remains a major cause of poor neurological outcome and increased mortality in the course of subarachnoid hemorrhage (Dorsch and King, 1994; Rabinstein et al., 2004; Macdonald et al., 2007).

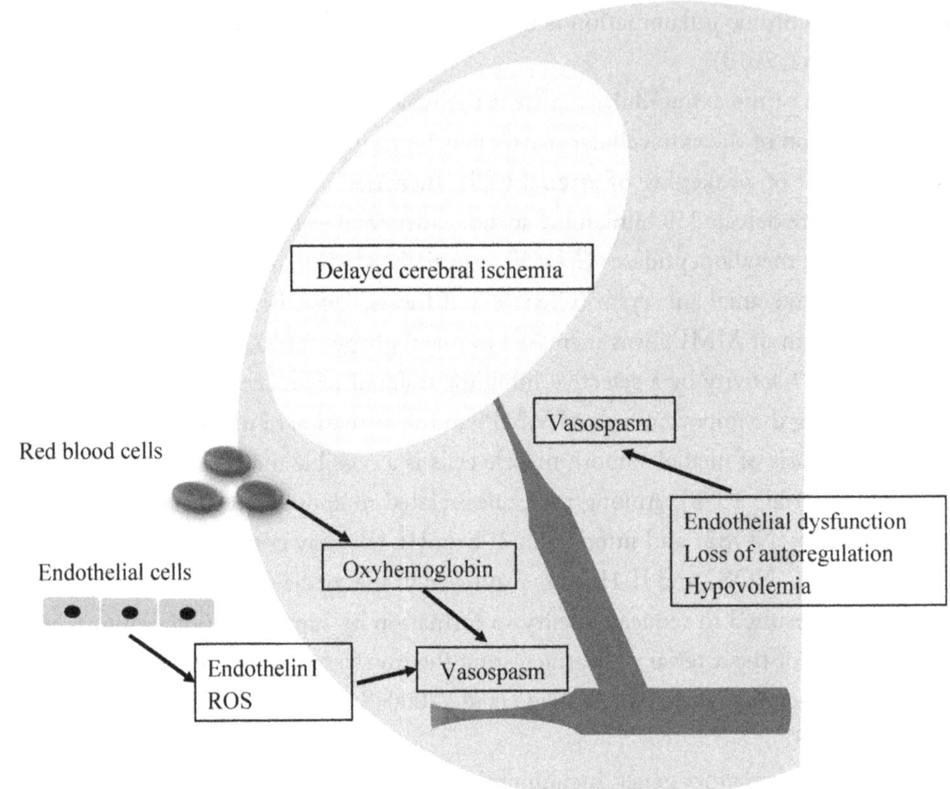

FIGURE 10: Proposed mechanisms of delayed cerebral ischemia after subarachnoid hemorrhage (Castanares-Zapatero and Hantson, 2011). Delayed cerebral ischemia is usually caused by a vasospasm. Several factors, including endothelial dysfunction, loss of autoregulation, and a hypovolemic component leading to a decrease in cerebral blood flow, may be involved in the etiology of vasospasm. The products derived from red blood cells, such as oxyhemoglobin, and from endothelial cells, such as endothelin 1 and reactive oxygen species, are mediators of the vasospasm.

Vasospasm occurs between the third and fifteenth day after the hemorrhage, with a peak at the tenth day. It is observed in 70% of patients and causes symptoms in 50% (Heros et al., 1983; Kassell et al., 1985; Vajkoczy et al., 2003). Vasospasm preferentially involves the vessels of the cranial base but also may affect small caliber vessels or diffusely the entire cerebral vascularization. The subsequent decrease in cerebral blood flow in the spastic arteries leads to delayed cerebral ischemia, which may develop into cerebral infarction (Vajkoczy et al., 2003; Suarez et al., 2006).

The etiology of vasospasm is complex and still poorly understood. Several factors have been shown to be involved, such as endothelial dysfunction, loss of autoregulation, and a hypovolemic component leading to a decrease in cerebral blood flow (Diringer, 2003; Dumont et al., 2003). At the acute phase, the presence of oxyhemoglobin in the subarachnoid spaces causes a local and systemic inflammatory reaction with activation of platelets and coagulation factors (Heros et al., 1983; Kassell et al., 1985; Yoshimoto et al., 2001). The products derived from red blood cells, such as oxyhemoglobin, and from endothelial cells, such as endothelin 1 and reactive oxygen species, are considered to be mediators of the vasospasm (Macdonald and Weir, 1991; Rubanyi, 1991). The poor prognosis of patients with delayed cerebral ischemia following subarachnoid hemorrhage remains a major issue responsible for death and infirmity. Although our understanding of the physiopathology of delayed cerebral ischemia and vasospasm has improved, patient outcome has not been significantly modified (Castanares-Zapatero and Hantson, 2011).

4.7 MOLECULAR GENETICS OF INTRACRANIAL ANEURYSM AND SUBARACHNOID HEMORRHAGE

Subarachnoid hemorrhage is 1.6 times more common in women than in men (Rinkel et al., 1998). Hormonal factors probably explain this sex-specific risk, given that it is higher in postmenopausal women than in premenopausal women (Longstreth et al., 1994). Smoking, alcohol consumption, and hypertension are also common risk factors for aneurysmal subarachnoid hemorrhage (Teunissen et al., 1996; Ruigrok et al., 2001). In addition to these environmental risk factors, genetic factors play an important role in the pathogenesis of subarachnoid hemorrhage associated with intracranial aneurysms. First-degree relatives of affected individuals are thus at up to seven times greater risk than is the general population (Bromberg et al., 1995; Schievink et al., 1995; Braekeleer et al., 1996; Ronkainen et al., 1997; Gaist et al., 2000), and ~10% of patients with aneurysmal subarachnoid hemorrhage have first- or second-degree relatives with subarachnoid hemorrhage or unruptured intracranial aneurysms (Norrgard et al., 1987; Ronkainen et al., 1993; Bromberg et al., 1995; Schievink et al., 1995; Wang et al., 1995; Braekeleer et al., 1996).

Several additional lines of evidence further support a role for genetic factors in the etiology of intracranial aneurysm. First, several genetic diseases, such as adult polycystic kidney disease (Chapman et al., 1992), Marfan syndrome (ter Berg et al., 1986), glucocorticoid-remediable aldosteronism (Litchfield et al., 1998), and Ehlers–Danlos syndrome type IV (de Paepe et al., 1988), increase the risk of the formation of intracranial aneurysm. Second, familial recurrence of nonsyndromic intracranial aneurysm has been described (Fox and Ko, 1980; Morooka and Waga, 1983; Maroun et al., 1986; Elshunnar and Whittle, 1990). Indeed, there is a three to fivefold increase in the risk for this condition in first-degree relatives of affected individuals compared with the general population (Stehbens, 1998; Ronkainen et al., 1999). Previous genetic linkage analyses have mapped loci for intracranial aneurysm to chromosomal regions 1p34.3–p36.13 (Nahed et al., 2005), 2p13 (Roos et al., 2004), 5q22–q31 (Onda et al., 2001), 7q11 (Onda et al., 2001), 11q24–q25 (Ozturk et al., 2006),

14q22 (Onda et al., 2001), 14q23–q31 (Ozturk et al., 2006), 17cen (Yamada et al., 2004), 19q13 (Yamada et al., 2004), 19q13.3 (Van der Voet et al., 2004), and Xp22 (Yamada et al., 2004).

Magnetic resonance angiography (MRI) is not sufficiently effective for screening the first-degree relatives of individuals with sporadic subarachnoid hemorrhage for intracranial aneurysms (The Magnetic Resonance Angiography in Relatives of Patients with Subarachnoid Hemorrhage Study Group 1999), and repeated screening is necessary to identify newly developed aneurysms in familial subarachnoid hemorrhage (Raaymakers et al., 1998). Given that a familial predisposition is the strongest risk factor for the development of intracranial aneurysms (Rinkel et al., 1998; Ruigrok et al., 2001), the identification of genetic risk factors might provide further diagnostic capability. In the future, genotype assessment might thus help to identify first-degree relatives of individuals with subarachnoid hemorrhage who are at high risk of developing one or more intracranial aneurysms (Ruigrok et al., 2005). Furthermore, identification of these genetic factors should provide insight into the pathophysiology of intracranial aneurysms (Ruigrok et al., 2005). The identification of disease susceptibility genes and increased understanding of the disease pathophysiology may lead to new therapeutic interventions to prevent the development, growth, or rupture of intracranial aneurysms (Ruigrok et al., 2005).

Although various candidate gene association studies of unrelated individuals have identified genes that are related to subarachnoid hemorrhage or intracranial aneurysm (Table 4), the genes that confer susceptibility to these conditions have not been determined definitively. Recent GWASs identified several loci and genes that confer susceptibility for intracranial aneurysm (Table 5). Candidate genes for intracranial aneurysm or subarachnoid hemorrhage that are of particular interest (*ELN*, *LIMK1*, *TNFRSF13B*, and five loci identified by GWASs) are reviewed in the following section.

4.7.1 Elastin Gene (*ELN*) and LIM Domain Kinase 1 Gene (*LIMK1*)

A functional haplotype spanning *ELN* and *LIMK1* was shown to confer susceptibility to intracranial aneurysm (Akagawa et al., 2006). *ELN* is located within the chromosome 7q11 linkage region and was recognized earlier to be a positional and functional candidate gene for intracranial aneurysm (Onda et al., 2001). However, allelic association studies yielded variable results (Hofer et al., 2003; Ruigrok et al., 2004). A systematic analysis of 166 SNPs and haplotypes that reside within the chromosome 7q linkage peak identified a highly significant association between intracranial aneurysm and a distinct linkage disequilibrium block containing the 3′ untranslated region of *ELN* and the promoter region of *LIMK1* (Akagawa et al., 2006). The strongest association was found with the *ELN* +695G→C tag SNP for a risk haplotype comprised of the functional *ELN* +502A insertion and the *LIMK1* −187C→T SNP. Both the genotype and haplotype associations

TABLE 4: Genes shown to be related to intracranial aneurysm or subarachnoid hemorrhage by linkage analyses or candidate gene association studies.

CHROMOSOMAL LOCUS	GENE NAME	GENE SYMBOL	REFERENCE
1p36.1	Heparan sulfate proteoglycan 2	*HSPG2*	Ruigrok et al. (2006a)
5q12–q14	Versican	*VCAN*	Ruigrok et al. (2006b)
5q23–q31	Fibrillin 2	*FBN2*	Ruigrok et al. (2006a)
6p21.3	Tumor necrosis factor	*TNF*	Yamada et al. (2006)
7p21	Interleukin 6	*IL6*	Morgan et al. (2006)
7q11.2	Elastin	*ELN*	Akagawa et al. (2006)
7q11.23	Lim domain kinase 1	*LIMK1*	Akagawa et al. (2006)
7q21.1	Cytochrome P450, family 3, subfamily A, polypeptide 4	*CYP3A4*	Yamada et al. (2008)
7q21.3–q22	Serpin peptidase inhibitor, clade E, member 1	*SERPINE1*	Ruigrok et al. (2006a)
7q22.1	Collagen, type I, alpha 2	*COL1A2*	Yoneyama et al. (2004)
7q36	Nitric oxide synthase 3	*NOS3*	Khurana et al. (2004)
9q34.1	Endoglin	*ENG*	Takenaka et al. (1999)
11q13	Uncoupling protein 3	*UCP3*	Yamada et al. (2006)
13q34	Collagen, type IV, alpha 1	*COL4A1*	Ruigrok et al. (2006a)
14q32.1	Serpin peptidase inhibitor, clade A, member 3	*SERPINA3*	Slowik et al. (2005)

TABLE 4: *(continued)*

CHROMOSOMAL LOCUS	GENE NAME	GENE SYMBOL	REFERENCE
16p13.3–p13.12	Polycystic kidney disease 1	*PKD1*	Watnick et al. (1999)
17p11.2	Tumor necrosis factor receptor superfamily, member 13B	*TNFRSF13B*	Inoue et al. (2006)
17q21.32	Integrin, beta 3	*ITGB3*	Iniesta et al. (2004)
17q23	Angiotensin I converting enzyme	*ACE*	Slowik et al. (2004a)
20q11.2–q13.1	Matrix metallo-peptidase 9	*MMP9*	Peters et al. (1999)
22q12	Heme oxygenase 1	*HMOX1*	Morgan et al. (2005)

were replicated in an independent cohort. Functional studies revealed that the *ELN* +502A insertion reduces the rate of *ELN* transcription, whereas the *LIMK1* −187C→T SNP reduces promoter activity (Dichgans and Hegele, 2007). Synergism between genetic variants of *ELN* and *LIMK1* in their effects on vascular stability and distensibility seems plausible because (i) elastin is a major structural component of the internal elastic lamina in cerebral arteries; (ii) *ELN* plays a key role in vascular development and remodeling; (iii) secreted elastin activates a G protein-coupled signaling pathway that stimulates organization of actin stress fibers; and (iv) LIMK1 is a regulator of the actin cytoskeleton (Dichgans and Hegele, 2007).

4.7.2 Tumor Necrosis Factor Receptor Superfamily, Member 13B Gene (*TNFRSF13B*)

Sequence variation in *TNFRSF13B* was shown to contribute to risk for intracranial aneurysm (Inoue et al., 2006). Sequence analysis of genes in a linkage peak on chromosome 17p revealed several potentially deleterious changes in *TNFRSF13B* that segregated with intracranial aneurysm in pedigrees. Sequencing of a portion of *TNFRSF13B* in a large case-control sample showed that several

TABLE 5: Chromosomal loci and genes shown to be related to intracranial aneurysm by genome-wide association studies.

CHROMOSOMAL LOCUS	DBSNP	NUCLEOTIDE SUBSTITUTION	GENE (NEARBY GENE)	REFERENCE
2q33.1	rs700651	G→A	*BOLL*	Bilguvar et al. (2008)
4q31.22	rs6842241	A→C	*EDNRA*	Low et al. (2012)
4q31.22	rs6841581	G→A	*EDNRA*	Low et al. (2012)
8q11.12–12.1	rs10958409	G→A	*SOX17*	Bilguvar et al. (2008)
8q12.1	rs9298506	A→G	*SOX17*	Yasuno et al. (2010)
9p21.3	rs1333040	C→T	*CDKN2B-AS1*	Bilguvar et al. (2008)
9p21.3	rs10757272	C→T	*CDKN2B-AS1*	Low et al. (2012)
10q24.32	rs12413409	G→A	*CNNM2*	Yasuno et al. (2010)
13q13.1	rs9315204	C→T	*STARD13*	Yasuno et al. (2010)
18q11.2	rs11661542	C→A	*RBBP8*	Yasuno et al. (2010)

potentially functional rare variants were more frequent in cases than in controls. Finally, association analyses suggested that one of the *TNFRSF13B* haplotypes was protective. Interactions of genetic factors such as *TNFRSF13B* with known risk factors for aneurysm formation, such as smoking and hypertension, remain an important area of research (Dichgans and Hegele, 2007).

4.7.3 Five Loci for Intracranial Aneurysm Identified by GWASs

A GWAS of intracranial aneurysm including Finnish, Dutch, and Japanese cohorts of 2196 cases and 8085 controls showed that common SNPs on 2q33.1, 8q11.12–12.1, and 9p21.3 were significantly associated with intracranial aneurysm (Bilguvar et al., 2008). In a follow-up GWAS, additional European cases and controls were included and the original Japanese replication cohort was increased, resulting in a cohort of 5891 cases and 14,181 controls (Yasuno et al., 2010). This follow-up study identified three new loci strongly associated with intracranial aneurysms on 18q11.2,

13q13.1, and 10q24.32. The previously discovered associations of 8q11.23–q12.1 and 9p21.3 were replicated (Yasuno et al., 2010).

The 8q11.23–q12.1 locus contains the SRY-box 17 gene (*SOX17*), which encodes a transcription factor that has a pivotal role in endothelial cell function (Francois et al., 2010). The strongly associated SNP was located in the 9p21.3 locus that contains *CDKN2B-AS1*. A recently described mutant mouse with a deletion corresponding to the human 9p21 locus showed a marked suppression of the expression of *CDKN2B* and *CDKN2A* (Visel et al., 2010). Aortic smooth muscle cells in culture from these mice showed increased proliferative activity compared with aortic smooth muscle cells from wild-type mice (Visel et al., 2010). The associated SNP on the 10q24.32 locus is located within the cyclin M2 gene (*CNNM2*). The 13q13.1 locus includes the StAR-related lipid transfer domain containing 13 gene (*STARD13*), of which overexpression leads to suppression in cell proliferation (Leung et al., 2005). The 18q11.2 locus contains the retinoblastoma binding protein 8 gene (*RBBP8*). RBBP8 is one of the proteins that bind directly to retinoblastoma protein, which regulates cell proliferation (Nevins, 1998). The gene products of the candidate genes in the five loci may be involved in cell proliferation (Ruigrok and Rinkel, 2010).

• • • •

CHAPTER 5

Clinical Implications

The increasing body of information garnered from studies on the genetics of stroke has resulted in the emergence of a greater understanding of the etiology of ischemic and hemorrhagic stroke. Such knowledge may have clinical implications for the prediction, diagnosis, prognosis, and treatment of stroke. The genes responsible for the pathogenesis of ischemic stroke, intracerebral hemorrhage, intracranial aneurysm, and subarachnoid hemorrhage as well as their encoded proteins are potentially important therapeutic targets in the design of new treatments for stroke. Genetic markers are potential diagnostic tools for assessment of individuals at risk of developing stroke. Genetic markers of stroke together with examination with MRI and MRA might also form the basis for promotion of preventive therapies in individuals at risk of stroke. It should remember, however, that gene–gene and gene–environment interactions make interpretation of information based on genetic markers of stroke more complex than is that of information based on markers for monogenic stroke. Another use of genetic markers might be to distinguish treatment responders from nonresponders (such as clopidogrel resistance) and to identify patients who are at risk of developing unfavorable side effects. It is likely that the use of gene polymorphisms to predict the response to and adverse effects of therapies for stroke will increase in the future and will give rise to major advances in patient care. Genetic analysis of strokes is thus likely to have important direct clinical applications.

• • • •

CHAPTER 6
Conclusion

I have summarized single-gene disorders associated with stroke and the candidate loci and genes for common forms of stroke. There has been a growing effort to find genetic variants that confer risk for stroke as a means to understand the underlying biological events. Clarification of the functional relevance of SNPs at various loci to ischemic stroke, intracerebral hemorrhage, and subarachnoid hemorrhage may provide insight into the pathogenesis of these conditions as well as into the role of genetic factors in their development. Such studies may ultimately lead to the personalized prevention of stroke. It may thus become possible to predict the future risk for stroke in each individual on the basis of conventional laboratory analyses, examination with MRI and MRA, and genetic analyses. Furthermore, it may be possible to prevent an individual from having stroke by medical intervention based on his or her genotypes for specific SNPs. In the future, we may have the ability to use specific agents particularized for certain genetic susceptibility factors, thereby increasing efficacy and limiting side effects of treatment. Identification of susceptibility genes for stroke and clarification of the functional relevance of genetic variants to this condition will thus contribute to the personalized prevention, early diagnosis, and treatment of stroke.

• • • •

References

Adams HP Jr, Bendixen BH, Kappelle LJ, Biller J, Love BB, Gordon DL, Marsh EE 3rd (1993). Classification of subtype of acute ischemic stroke. Definitions for use in a multicenter clinical trial. TOAST. Trial of Org 10172 in Acute Stroke Treatment. *Stroke* 24: pp. 35–41.

Akagawa H, Tajima A, Sakamoto Y, Krischek B, Yoneyama T, Kasuya H, Onda H, Hori T, Kubota M, Machida T, Saeki N, Hata A, Hashiguchi K, Kimura E, Kim CJ, Yang TK, Lee JY, Kimm K, Inoue I (2006). A haplotype spanning two genes, ELN and LIMK1, decreases their transcripts and confers susceptibility to intracranial aneurysms. *Hum Mol Genet* 15: pp. 1722–34.

Alamowitch S, Plaisier E, Favrole P, Prost C, Chen Z, Van Agtmael T, Marro B, Ronco P (2009). Cerebrovascular disease related to COL4A1 mutations in HANAC syndrome. *Neurology* 73: pp. 1873–82.

Alberts MJ (2003). Stroke genetics update. *Stroke* 34: pp. 342–4.

Alberts MJ, Davis JP, Graffagnino C, McClenny C, Delong D, Granger C, Herbstreith MH, Boteva K, Marchuk DA, Roses AD (1997). Endoglin gene polymorphism as a risk factor for sporadic intracerebral hemorrhage. *Ann Neurol* 41: pp. 683–6.

Alberts MJ, McCarron MO, Hoffmann KL, Graffagnino C (2002). Familial clustering of intracerebral hemorrhage: a prospective study in North Carolina. *Neuroepidemiology* 21: pp. 18–21.

Anderson CD, Biffi A, Rost NS, Cortellini L, Furie KL, Rosand J (2010). Chromosome 9p21 in ischemic stroke: population structure and meta-analysis. *Stroke* 41: pp. 1123–31.

Aoki T, Kataoka H, Ishibashi R, Nozaki K, Egashira K, Hashimoto N (2009). Impact of monocyte chemoattractant protein-1 deficiency on cerebral aneurysm formation. *Stroke* 40: pp. 942–51.

Aoki T, Kataoka H, Morimoto M, Nozaki K, Hashimoto N (2007). Macrophage-derived matrix metalloproteinase-2 and -9 promote the progression of cerebral aneurysms in rats. *Stroke* 38: pp. 162–9.

Aoki T, Nishimura M (2010). Molecular mechanism of cerebral aneurysm formation focusing on NF-κB as a key mediator of inflammation. *J Biorheol* 24: pp. 16–21.

Ardizzone TD, Lu A, Wagner KR, Tang Y, Ran R, Sharp FR (2004). Glutamate receptor blockade attenuates glucose hypermetabolism in perihematomal brain after experimental intracerebral hemorrhage in rat. *Stroke* 35: pp. 2587–91.

Aronowski J, Hall CE (2005). New horizons for primary intracerebral hemorrhage treatment: experience from preclinical studies. *Neurol Res* 27: pp. 268–79.

Aronowski J, Zhao X (2011). Molecular pathophysiology of cerebral hemorrhage: secondary brain injury. *Stroke* 42: pp. 1781–6.

Bak S, Gaist D, Sindrup SH, Skytthe A, Christensen K (2002). Genetic liability in stroke: a long-term follow-up study of Danish twins. *Stroke* 33: pp. 769–74.

Baker RI, Eikelboom J, Lofthouse E, Staples N, Afshar-Kharghan V, Lopez JA, Shen Y, Berndt MC, Hankey G (2001). Platelet glycoprotein Ibα Kozak polymorphism is associated with an increased risk of ischemic stroke. *Blood* 98: pp. 36–40.

Ballermann BJ, Dardik A, Eng E, Liu A (1998). Shear stress and the endothelium. *Kidney Int* 67(Suppl.): pp. S100–8.

Bevan S, Dichgans M, Gschwendtner A, Kuhlenbäumer G, Ringelstein EB, Markus HS (2008). Variation in the PDE4D gene and ischemic stroke risk. A systematic review and meta-analysis on 5200 cases and 6600 controls. *Stroke* 39: pp. 1966–71.

Bevan S, Porteous L, Sitzer M, Markus HS (2005). Phosphodiesterase 4D gene, ischemic stroke, and asymptomatic carotid atherosclerosis. *Stroke* 36: pp. 949–53.

Bilguvar K, Yasuno K, Niemelä M, Ruigrok YM, von Und Zu Fraunberg M, van Duijn CM, van den Berg LH, Mane S, Mason CE, Choi M, Gaál E, Bayri Y, Kolb L, Arlier Z, Ravuri S, Ronkainen A, Tajima A, Laakso A, Hata A, Kasuya H, Koivisto T, Rinne J, Ohman J, Breteler MM, Wijmenga C, State MW, Rinkel GJ, Hernesniemi J, Jääskeläinen JE, Palotie A, Inoue I, Lifton RP, Günel M (2008). Susceptibility loci for intracranial aneurysm in European and Japanese populations. *Nat Genet* 40: pp. 1472–7.

Bonventre JV, Huang Z, Taheri MR, O'Leary E, Li E, Moskowitz MA, Sapirstein A (1997). Reduced fertility and postischaemic brain injury in mice deficient in cytosolic phospholipase A2. *Nature* 390: pp. 622–5.

Braekeleer DM, Pérussee L, Cantin L, Bouchard JM, Mathieu J (1996). A study of inbreeding and kinship in intracranial aneurysms in the Saguenay Lac-Saint-Jean region (Quebec, Canada). *Ann Hum Genet* 60: pp. 99–104.

Brass LM, Isaacsohn JL, Merikangas KR, Robinette CD (1992). A study of twins and stroke. *Stroke* 23: pp. 221–3.

Bromberg JE, Rinkel GJ, Algra A, Greebe P, van Duyn CM, Hasan D, Limburg M, ter Berg HW, Wijdicks EF, van Gijn J (1995). Subarachnoid haemorrhage in first and second degree relatives of patients with subarachnoid haemorrhage. *BMJ* 311: pp. 288–9.

Broughton BRS, Reutens DC, Sobey CG (2009). Apoptotic mechanisms after cerebral Ischemia. *Stroke* 40: pp. e331–9.

Bruno G, Todor R, Lewis I, Chyatte D (1998). Vascular extracellular matrix remodeling in cerebral aneurysms. *J Neurosurg* 89: pp. 431–40.

Carlsson LE, Santoso S, Spitzer C, Kessler C, Greinacher A (1999). The α_2 gene coding sequence T807/A873 of the platelet collagen receptor integrin $\alpha_2\beta_1$ might be a genetic risk factor for the development of stroke in younger patients. *Blood* 93: pp. 3583–6.

Carlsson M, Orho-Melander M, Hedenbro J, Almgren P, Groop LC. The T 54 allele of the intestinal fatty acid-binding protein 2 is associated with a parental history of stroke (2000). *J Clin Endocrinol Metab* 85: pp. 2801–4.

Carr FJ, McBride MW, Carswell HV, Graham D, Strahorn P, Clark JS, Charchar FJ, Dominiczak AF (2002). Genetic aspects of stroke: human and experimental studies. *J Cerebr Blood Flow Metab* 22: pp. 767–73.

Castanares-Zapatero D, Hantson P (2011). Pharmacological treatment of delayed cerebral ischemia and vasospasm in subarachnoid hemorrhage. *Ann Intensive Care* 1: p. 12.

Chabriat H, Vahedi K, Iba-Zizen MT, Joutel A, Nibbio A, Nagy TG, Krebs MO, Julien J, Dubois B, Ducrocq X, Levasseur M, Homeyer P, Mas JL, Lyon-Caen O, Tournier Lasserve E, Bousser MG (1995). Clinical spectrum of CADASIL: a study of 7 families. Cerebral autosomal dominant arteriopathy with subcortical infarcts and leukoencephalopathy. *Lancet* 346: pp. 934-939.

Chapman AB, Rubinstein D, Hughes, R, Stears JC, Earnest MP, Johnson AM, Gabow PA, Kaehny WD (1992). Intracranial aneurysms in autosomal dominant polycystic kidney disease. *N Engl J Med* 327: pp. 916–20.

Chyatte D, Bruno G, Desai S, Todor DR (1999). Inflammation and intracranial aneurysms. *Neurosurgery* 45: pp. 1137–46.

Chyatte D, Reilly J, Tilson MD (1990). Morphometric analysis of reticular and elastin fibers in the cerebral arteries of patients with intracranial aneurysms. *Neurosurgery* 26: pp. 939–43.

Ciafaloni E, Ricci E, Shanske S, Moraes CT, Silvestri G, Hirano M, Simonetti S, Angelini C, Donati MA, Garcia C, Martinuzzi A, Mosewich R, Servidei S, Zammarchi E, Bonilla E, DeVivo DC, Rowland LP, Schon EA, DiMauro S (1992). MELAS: clinical features, biochemistry, and molecular genetics. *Ann Neurol* 31: pp. 391-398.

Cipollone F, Toniato E, Martinotti S, Fazia M, Iezzi A, Cuccurullo C, Pini B, Ursi S, Vitullo G, Averna M, Arca M, Montali A, Campagna F, Ucchino S, Spigonardo F, Taddei S, Virdis A, Ciabattoni G, Notarbartolo A, Cuccurullo F, Mezzetti A; Identification of New Elements of Plaque Stability (INES) Study Group (2004). A polymorphism in the cyclooxygenase 2 gene as an inherited protective factor against myocardial infarction and stroke. *JAMA* 291: pp. 2221–8.

Cole JW, Meschia JF (2011). Stroke Genetics Update: 2011. *Curr Cardiovasc Risk Rep* 5: pp. 533–41.

Cole FM, Yates PO (1967). Pseudo-aneurysms in relationship to massive cerebral hemorrhage. *J Neurol Neurosurg Psychiatry* 30: pp. 61–6.

Crutchfield KE, Patronas NJ, Dambrosia JM, Frei KP, Banerjee TK, Barton NW, Schiffmann R (1998). Quantitative analysis of cerebral vasculopathy in patients with Fabry disease. *Neurology* 50: pp. 1746–9.

Chyatte D, Lewis I (1997). Gelatinase activity and the occurrence of cerebral aneurysms. *Stroke* 28: pp. 799–804.

Davis ME, Grumbach IM, Fukai T, Cutchins A, Harrison DG (2004). Shear stress regulates endothelial nitric-oxide synthase promoter activity through nuclear factor kappaB binding. *J Biol Chem* 279: pp. 163–8.

de Paepe A, van Landegem W, de Keyser F, de Reuck J (1988). Association of multiple intracranial aneurysms and collagen type III deficiency. *Clin Neurol Neurosurg* 90: pp. 53–6.

de Rooij NK, Linn FH, van der Plas JA, Algra A, Rinkel GJ (2007). Incidence of subarachnoid haemorrhage: a systematic review with emphasis on region, age, gender and time trends. *J Neurol Neurosurg Psychiatry* 78: pp. 1365–72.

De Stefano V, Chiusolo P, Paciaroni K, Casorelli I, Rossi E, Molinari M, Servidei S, Tonali PA, Leone G (1998). Prothrombin G20210A mutant genotype is a risk factor for cerebrovascular ischemic disease in young patients. *Blood* 91: pp. 3562–5.

Debette S, Bis JC, Fornage M, Schmidt H, Ikram MA, Sigurdsson S, Heiss G, Struchalin M, Smith AV, van der Lugt A, DeCarli C, Lumley T, Knopman DS, Enzinger C, Eiriksdottir G, Koudstaal PJ, DeStefano AL, Psaty BM, Dufouil C, Catellier DJ, Fazekas F, Aspelund T, Aulchenko YS, Beiser A, Rotter JI, Tzourio C, Shibata DK, Tscherner M, Harris TB, Rivadeneira F, Atwood LD, Rice K, Gottesman RF, van Buchem MA, Uitterlinden AG, Kelly-Hayes M, Cushman M, Zhu Y, Boerwinkle E, Gudnason V, Hofman A, Romero JR, Lopez O, van Duijn CM, Au R, Heckbert SR, Wolf PA, Mosley TH, Seshadri S, Breteler MM, Schmidt R, Launer LJ, Longstreth WT Jr (2010). Genome-wide association studies of MRI-defined brain infarcts: meta-analysis from the CHARGE Consortium. *Stroke* 41: pp. 210–7.

Dichgans M, Hegele RA (2007). Update on the genetics of stroke and cerebrovascular disease 2006. *Stroke* 38: pp. 216–8.

Diringer MN (2003). Subarachnoid hemorrhage: a multiple-organ system disease. *Crit Care Med* 31: pp. 1884–5.

Dirnagl U, Iadecola C, Moskowitz MA (1999). Pathobiology of ischaemic stroke: an integrated view. *Trends Neurosci* 22: pp. 391–7.

Dixon RA, Diehl RE, Opas E, Rands E, Vickers PJ, Evans JF, Gillard JW, Miller DK (1990). Requirement of a 5-lipoxygenase-activating protein for leukotriene synthesis. *Nature* 343: pp. 282–4.

Dorsch NW, King MT (1994). A review of cerebral vasospasm in aneurismal subarachnoid haemorrhage Part I: Incidence and effects. *J Clin Neurosci* 1: pp. 19–26.

Dumont AS, Dumont RJ, Chow MM, Lin CL, Calisaneller T, Ley KF, Kassell NF, Lee KS (2003). Cerebral vasospasm after subarachnoid hemorrhage: putative role of inflammation. *Neurosurgery* 53: pp. 123–33.

Elbaz A, Poirier O, Canaple S, Chedru F, Cambien F, Amarenco P (2000a). The association between the Val34Leu polymorphism in the factor XIII gene and brain infarction. *Blood* 95: pp. 586–91.

Elbaz A, Poirier O, Moulin T, Chedru F, Cambien F, Amarenco P (2000b). Association between the Glu298Asp polymorphism in the endothelial constitutive nitric oxide synthase gene and brain infarction. The GENIC Investigators. *Stroke* 31: pp. 1634–9.

Elshunnar KS, Whittle IR (1990). Familial intracranial aneurysms: report of five families. *Br J Neurosurg* 4: pp. 181–6.

Enter C, Muller-Hocker J, Zierz S, Kurlemann G, Pongratz D, Forster C, Obermaier-Kusser B, Gerbitz KD (1991). A specific point mutation in the mitochondrial genome of Caucasians with MELAS. *Hum Genet* 88: pp. 233–6.

Felberg RA, Grotta JC, Shirzadi AL, Strong R, Narayana P, Hill-Felberg SJ, Aronowski J (2002). Cell death in experimental intracerebral hemorrhage: the "black hole" model of hemorrhagic damage. *Ann Neurol* 51: pp. 517–24.

Fergusen S, Macdonald RL (2007). Predictors of cerebral infarction in patients with aneurysmal subarachnoid hemorrhage. *Neurosurgery* 60: pp. 658–67.

Floßmann E, Schulz UGR, Rothwell PM (2004). Systematic review of methods and results of studies of the genetic epidemiology of ischemic stroke. *Stroke* 35: pp. 212–27.

Fornage M, Lee CR, Doris PA, Bray MS, Heiss G, Zeldin DC, Boerwinkle E (2005). The soluble epoxide hydrolase gene harbors sequence variation associated with susceptibility to and protection from incident ischemic stroke. *Hum Mol Genet* 14: pp. 2829–37.

Fox JL, Ko JP (1980). Familial intracranial aneurysms. Six cases among 13 siblings. *J Neurosurg* 52: pp. 501–3.

Francois M, Koopman P, Beltrame M (2010). SoxF genes: key players in the development of the cardio-vascular system. *Int J Biochem Cell Biol* 42: pp. 445–8.

Frikke-Schmidt R, Nordestgaard BG, Schnohr P, Tybjaerg-Hansen A (2004). Single nucleotide polymorphism in the low-density lipoprotein receptor is associated with a threefold risk of stroke. A case-control and prospective study. *Eur Heart J* 25: pp. 943–51.

Fukuda S, Hashimoto N, Naritomi H, Nagata I, Nozaki K, Kondo S, Kurino M, Kikuchi H (2000). Prevention of rat cerebral aneurysm formation by inhibition of nitric oxide synthase. *Circulation* 101: pp. 2532–8.

Fukumoto S, Koyama H, Hosoi M, Yamakawa K, Tanaka S, Morii H, Nishizawa Y (1999). Distinct role of cAMP and cGMP in the cell cycle control of vascular smooth muscle cells: cGMP delays cell cycle transition through suppression of cyclin D1 and cyclin-dependent kinase 4 activation. *Circ Res* 85: pp. 985–91.

Gaetani P, Tartara F, Grazioli V, Tancioni F, Infuso L, Baena R (1998). Collagen cross-linkage, elastolytic and collagenolytic activities in cerebral aneurysms: a preliminary investigation. *Life Sci* 63: pp. 285–92.

Gaetani P, Tartara F, Tancioni F, Baena R, Casari E, Alfano M, Grazioli V (1997). Deficiency of total collagen content and of deoxypyridinoline in intracranial aneurysm walls. *FEBS Lett* 404: pp. 303–6.

Gaist D, Vaeth M, Tsiropulos I, Christensen K, Corder E, Olsen J, Sorensen HT (2000). Risk of subarachnoid haemorrhage in first degree relatives of patients with subarachnoid haemorrhage: follow up study based on national registries in Denmark. *BMJ* 320: pp. 141–5.

Ginsberg MD (1997). The new language of cerebral ischemia. *Am J Neuroradiol* 18: pp. 1435–45.

Goldstein LB, Adams R, Becker K, Furberg CD, Gorelick PB, Hademenos G, Hill M, Howard G, Howard VJ, Jacobs B, Levine SR, Mosca L, Sacco RL, Sherman DG, Wolf PA, del Zoppo GJ (2001). Primary prevention of ischemic stroke: a statement for healthcare professionals from the Stroke Council of the American Heart Association. *Stroke* 32: pp. 280–99.

Gong C, Hoff JT, Keep RF (2000). Acute inflammatory reaction following experimental intracerebral hemorrhage in rat. *Brain Res* 871: pp. 57–65.

Gould DB, Phalan FC, Breedveld GJ, van Mil SE, Smith RS, Schimenti JC, Aguglia U, van der Knaap MS, Heutink P, John SW (2005). Mutations in Col4a1 cause perinatal cerebral hemorrhage and porencephaly. *Science* 308: pp. 1167–71.

Gould DB, Phalan FC, van Mil SE, Sundberg JP, Vahedi K, Massin P, Bousser MG, Heutink P, Miner JH, Tournier-Lasserve E, John SW (2006). Role of COL4A1 in small-vessel disease and hemorrhagic stroke. *N Engl J Med* 354: pp. 1489–96.

Greenberg SM, Briggs ME, Hyman BT, Kokoris GJ, Takis C, Kanter DS, Kase CS, Pessin MS (1996). Apolipoprotein E epsilon 4 is associated with the presence and earlier onset of hemorrhage in cerebral amyloid angiopathy. *Stroke* 27: pp. 1333–7.

Greenberg SM, Rebeck GW, Vonsattel JP, Gomez-Isla T, Hyman BT (1995). Apolipoprotein E epsilon 4 and cerebral hemorrhage associated with amyloid angiopathy. *Ann Neurol* 38: pp. 254–9.

Greenberg SM, Vonsattel JP, Segal AZ, Chiu RI, Clatworthy AE, Liao A, Hyman BT, Rebeck GW (1998). Association of apolipoprotein E epsilon 2 and vasculopathy in cerebral amyloid angiopathy. *Neurology* 50: pp. 961–5.

Gretarsdottir S, Sveinbjornsdottir S, Jonsson HH, Jakobsson F, Einarsdottir E, Agnarsson U, Shkolny D, Einarsson G, Gudjonsdottir HM, Valdimarsson EM, Einarsson OB, Thorgeirsson G, Hadzic R, Jonsdottir S, Reynisdottir ST, Bjarnadottir SM, Gudmundsdottir T, Gudlaugsdottir GJ, Gill R, Lindpaintner K, Sainz J, Hannesson HH, Sigurdsson GT, Frigge ML, Kong A, Gudnason V, Stefansson K, Gulcher JR (2002). Localization of a susceptibility gene for common forms of stroke to 5q12. *Am J Hum Genet* 70: pp. 593–603.

Gretarsdottir S, Thorleifsson G, Manolescu A, Styrkarsdottir U, Helgadottir A, Gschwendtner A, Kostulas K, Kuhlenbäumer G, Bevan S, Jonsdottir T, Bjarnason H, Saemundsdottir J, Palsson S, Arnar DO, Holm H, Thorgeirsson G, Valdimarsson EM, Sveinbjörnsdottir S, Gieger C, Berger K, Wichmann HE, Hillert J, Markus H, Gulcher JR, Ringelstein EB, Kong A, Dichgans M, Gudbjartsson DF, Thorsteinsdottir U, Stefansson K (2008). Risk variants for atrial fibrillation on chromosome 4q25 associate with ischemic stroke. *Ann Neurol* 64: pp. 402–9.

Gretarsdottir S, Thorleifsson G, Reynisdottir ST, Manolescu A, Jonsdottir S, Jonsdottir T, Gudmundsdottir T, Bjarnadottir SM, Einarsson OB, Gudjonsdottir HM, Hawkins M, Gudmundsson G, Gudmundsdottir H, Andrason H, Gudmundsdottir AS, Sigurdardottir M, Chou TT, Nahmias J, Goss S, Sveinbjornsdottir S, Valdimarsson EM, Jakobsson F, Agnarsson U, Gudnason V, Thorgeirsson G, Fingerle J, Gurney M, Gudbjartsson D, Frigge ML, Kong A, Stefansson K, Gulcher JR (2003). The gene encoding phosphodiesterase 4D confers risk of ischemic stroke. *Nat Genet* 35: pp. 131–8.

Gschwendtner A, Bevan S, Cole JW, Plourde A, Matarin M, Ross-Adams H, Meitinger T, Wichmann E, Mitchell BD, Furie K, Slowik A, Rich SS, Syme PD, MacLeod MJ, Meschia JF, Rosand J, Kittner SJ, Markus HS, Müller-Myhsok B, Dichgans M; International Stroke Genetics Consortium (2009). Sequence variants on chromosome 9p21.3 confer risk for atherosclerotic stroke. *Ann Neurol* 65: pp. 531–9.

Gudbjartsson DF, Holm H, Gretarsdottir S, Thorleifsson G, Walters GB, Thorgeirsson G, Gulcher J, Mathiesen EB, Njølstad I, Nyrnes A, Wilsgaard T, Hald EM, Hveem K, Stoltenberg C, Kucera G, Stubblefield T, Carter S, Roden D, Ng MC, Baum L, So WY, Wong KS, Chan JC, Gieger C, Wichmann HE, Gschwendtner A, Dichgans M, Kuhlenbäumer G, Berger K, Ringelstein EB, Bevan S, Markus HS, Kostulas K, Hillert J, Sveinbjörnsdóttir S, Valdimarsson EM, Løchen ML, Ma RC, Darbar D, Kong A, Arnar DO, Thorsteinsdottir U, Stefansson K (2009). A sequence variant in ZFHX3 on 16q22 associates with atrial fibrillation and ischemic stroke. *Nat Genet* 41: pp. 876–8.

Gulcher JR, Gretarsdottir S, Helgadottir A, Stefansson K (2005). Genes contributing to risk for common forms of stroke. *Trends Mol Med* 11: pp. 217–24.

Guo F, Li Z, Song L, Han T, Feng Q, Guo Y, Xu J, He M, You C (2007). Increased apoptosis and cysteinyl aspartate specific protease-3 gene expression in human intracranial aneurysm. *J Clin Neurosci* 14: pp. 550–5.

Hademenos GJ, Alberts MJ, Awad I, Mayberg M, Shepard T, Jagoda A, Latchaw RE, Todd HW, Viste K, Starke R, Girgus MS, Marler J, Emr M, Hart N (2001). Advances in the genetics of cerebrovascular disease and stroke. *Neurology* 56: pp. 997–1008.

Hara K, Shiga A, Fukutake T, Nozaki H, Miyashita A, Yokoseki A, Kawata H, Koyama A, Arima K, Takahashi T, Ikeda M, Shiota H, Tamura M, Shimoe Y, Hirayama M, Arisato T, Yanagawa S, Tanaka A, Nakano I, Ikeda S, Yoshida Y, Yamamoto T, Ikeuchi T, Kuwano R, Nishizawa M, Tsuji S, Onodera O (2009). Association of HTRA1 mutations and familial ischemic cerebral small-vessel disease. *N Engl J Med* 360: pp. 1729–39.

Harismendy O, Notani D, Song X, Rahim NG, Tanasa B, Heintzman N, Ren B, Fu XD, Topol EJ, Rosenfeld MG, Frazer KA (2011). 9p21 DNA variants associated with coronary artery disease impair interferon-γ signalling response. *Nature* 470: pp. 264–8.

Hassan A, Markus HS (2000). Genetics and ischaemic stroke. *Brain* 123: pp. 1784–812.

Hata J, Matsuda K, Ninomiya T, Yonemoto K, Matsushita T, Ohnishi Y, Saito S, Kitazono T, Ibayashi S, Iida M, Kiyohara Y, Nakamura Y, Kubo M (2007). Functional SNP in an Sp1-binding site of AGTRL1 gene is associated with susceptibility to brain infarction. *Hum Mol Genet* 16: pp. 630–9.

Havasi V, Szolnoki Z, Talian G, Bene J, Komlosi K, Maasz A, Somogyvari F, Kondacs A, Szabo M, Fodor L, Bodor A, Melegh B (2006). Apolipoprotein A5 gene promoter region T–1131C polymorphism associates with elevated circulating triglyceride levels and confers susceptibility for development of ischemic stroke. *J Mol Neurosci* 29: pp. 177–83.

Hegedus K (1984). Some observations on reticular fibers in the media of the major cerebral arteries: a comparative study of patients without vascular diseases and those with ruptured berry aneurysms. *Surg Neurol* 22: pp. 301–7.

Hegener HH, Diehl KA, Kurth T, Gaziano JM, Ridker PM, Zee RY (2006). Polymorphisms of prostaglandin-endoperoxide synthase 2 gene, and prostaglandin-E receptor 2 gene, C-reactive protein concentrations and risk of atherothrombosis: a nested case-control approach. *J Thromb Haemost* 4: pp. 1718–22.

Helgadottir A, Gretarsdottir S, St Clair D, Manolescu A, Cheung J, Thorleifsson G, Pasdar A, Grant SF, Whalley LJ, Hakonarson H, Thorsteinsdottir U, Kong A, Gulcher J, Stefansson K, MacLeod MJ (2005). Association between the gene encoding 5-lipoxygenase-activating protein and stroke replicated in a Scottish population. *Am J Hum Genet* 76: pp. 505–9.

Helgadottir A, Manolescu A, Thorleifsson G, Gretarsdottir S, Jonsdottir H, Thorsteinsdottir U, Samani NJ, Gudmundsson G, Grant SF, Thorgeirsson G, Sveinbjornsdottir S, Valdimarsson EM, Matthiasson SE, Johannsson H, Gudmundsdottir O, Gurney ME, Sainz J, Thorhallsdottir M, Andresdottir M, Frigge ML, Topol EJ, Kong A, Gudnason V, Hakonarson H,

Gulcher JR, Stefansson K (2004). The gene encoding 5-lipoxygenase activating protein confers risk of myocardial infarction and stroke. *Nat Genet* 36: pp. 233–9.

Helgadottir A, Thorleifsson G, Manolescu A, Gretarsdottir S, Blondal T, Jonasdottir A, Jonasdottir A, Sigurdsson A, Baker A, Palsson A, Masson G, Gudbjartsson DF, Magnusson KP, Andersen K, Levey AI, Backman VM, Matthiasdottir S, Jonsdottir T, Palsson S, Einarsdottir H, Gunnarsdottir S, Gylfason A, Vaccarino V, Hooper WC, Reilly MP, Granger CB, Austin H, Rader DJ, Shah SH, Quyyumi AA, Gulcher JR, Thorgeirsson G, Thorsteinsdottir U, Kong A, Stefansson K (2007). A common variant on chromosome 9p21 affects the risk of myocardial infarction. *Science* 316: pp. 1491–3.

Helgadottir A, Thorleifsson G, Magnusson KP, Grétarsdottir S, Steinthorsdottir V, Manolescu A, Jones GT, Rinkel GJ, Blankensteijn JD, Ronkainen A, Jääskeläinen JE, Kyo Y, Lenk GM, Sakalihasan N, Kostulas K, Gottsäter A, Flex A, Stefansson H, Hansen T, Andersen G, Weinsheimer S, Borch-Johnsen K, Jorgensen T, Shah SH, Quyyumi AA, Granger CB, Reilly MP, Austin H, Levey AI, Vaccarino V, Palsdottir E, Walters GB, Jonsdottir T, Snorradottir S, Magnusdottir D, Gudmundsson G, Ferrell RE, Sveinbjornsdottir S, Hernesniemi J, Niemelä M, Limet R, Andersen K, Sigurdsson G, Benediktsson R, Verhoeven EL, Teijink JA, Grobbee DE, Rader DJ, Collier DA, Pedersen O, Pola R, Hillert J, Lindblad B, Valdimarsson EM, Magnadottir HB, Wijmenga C, Tromp G, Baas AF, Ruigrok YM, van Rij AM, Kuivaniemi H, Powell JT, Matthiasson SE, Gulcher JR, Thorgeirsson G, Kong A, Thorsteinsdottir U, Stefansson K (2008). The same sequence variant on 9p21 associates with myocardial infarction, abdominal aortic aneurysm and intracranial aneurysm. *Nat Genet* 40: pp. 217–24.

Heros RC, Zervas NT, Varsos V (1983). Cerebral vasospasm after subarachnoid hemorrhage: an update. *Ann Neurol* 14: pp. 599–608.

Herrmann SM, Funke-Kaiser H, Schmidt-Petersen K, Nicaud V, Gautier-Bertrand M, Evans A, Kee F, Arveiler D, Morrison C, Orzechowski HD, Elbaz A, Amarenco P, Cambien F, Paul M (2001). Characterization of polymorphic structure of cathepsin G gene: role in cardiovascular and cerebrovascular diseases. *Arterioscler Thromb Vasc Biol* 21: pp. 1538–43.

Hickenbottom SL, Grotta JC, Strong R, Denner LA, Aronowski J (1999). Nuclear factor-kappa B and cell death after experimental intracerebral hemorrhage in rats. *Stroke* 30: pp. 2472–7.

Hofer A, Hermans M, Kubassek N, Sitzer M, Funke H, Stogbauer F, Ivaskevicius V, Oldenburg J, Burtscher J, Knopp U, Schoch B, Wanke I, Hubner F, Deinsberger W, Meyer B, Boecher-Schwarz H, Poewe W, Raabe A, Steinmetz H, Auburger G (2003). Elastin polymorphism haplotype and intracranial aneurysms are not associated in Central Europe. *Stroke* 34: pp. 1207–11.

Hop JW, Rinkel GJE, Algra A, van Gijn J (1997). Case fatality rates and functional outcome after subarachnoid haemorrhage: a systematic review. *Stroke* 28: pp. 660–4.

Houslay MD, Adams DR (2003). PDE4 cAMP phosphodiesterases: modular enzymes that orchestrate signalling cross-talk, desensitization and compartmentalization. *Biochem J* 370: pp. 1–18.

Hsieh K, Lalouschek W, Schillinger M, Endler G, Reisinger M, Janisiw M, Lang W, Cheng S, Wagner O, Mannhalter C (2005). Impact of αENaC polymorphisms on the risk of ischemic cerebrovascular events: a multicenter case-control study. *Clin Chem* 51: pp. 952–6.

Huang FP, Xi G, Keep RF, Hua Y, Nemoianu A, Hoff JT (2002). Brain edema after experimental intracerebral hemorrhage: role of hemoglobin degradation products. *J Neurosurg* 96: pp. 287–93.

Humphries SE, Morgan L (2004). Genetic risk factors for stroke and carotid atherosclerosis: insights into pathophysiology from candidate gene approaches. *Lancet Neurol* 3: pp. 227–35.

Iacoviello L, Di Castelnuovo A, Gattone M, Pezzini A, Assanelli D, Lorenzet R, Del Zotto E, Colombo M, Napoleone E, Amore C, D'Orazio A, Padovani A, de Gaetano G, Giannuzzi P, Donati MB; IGIGI Investigators (2005). Polymorphisms of the interleukin-1β gene affect the risk of myocardial infarction and ischemic stroke at young age and the response of mononuclear cells to stimulation in vitro. *Arterioscler Thromb Vasc Biol* 25: pp. 222–7.

Ikram MA, Seshadri S, Bis JC, Fornage M, DeStefano AL, Aulchenko YS, Debette S, Lumley T, Folsom AR, van den Herik EG, Bos MJ, Beiser A, Cushman M, Launer LJ, Shahar E, Struchalin M, Du Y, Glazer NL, Rosamond WD, Rivadeneira F, Kelly-Hayes M, Lopez OL, Coresh J, Hofman A, DeCarli C, Heckbert SR, Koudstaal PJ, Yang Q, Smith NL, Kase CS, Rice K, Haritunians T, Roks G, de Kort PL, Taylor KD, de Lau LM, Oostra BA, Uitterlinden AG, Rotter JI, Boerwinkle E, Psaty BM, Mosley TH, van Duijn CM, Breteler MM, Longstreth WT Jr, Wolf PA (2009). Genomewide association studies of stroke. *N Engl J Med* 360: pp. 1718–28.

Inagawa T, Tokuda Y, Ohbayashi N, Takaya M, Moritake K (1995). Study of aneurysmal subarachnoid hemorrhage in Izumo City, Japan. *Stroke* 26: pp. 761–6.

Indolfi C, Avvedimento EV, Di Lorenzo E, Esposito G, Rapacciuolo A, Giuliano P, Grieco D, Cavuto L, Stingone AM, Ciullo I, Condorelli G, Chiariello M (1997). Activation of cAMP–PKA signaling *in vivo* inhibits smooth muscle cell proliferation induced by vascular injury. *Nat Med* 3: pp. 775–9.

Indolfi C, Di Lorenzo E, Rapacciuolo A, Stingone AM, Stabile E, Leccia A, Torella D, Caputo R, Ciardiello F, Tortora G, Chiariello M (2000). 8-Chloro-cAMP inhibits smooth muscle cell proliferation *in vitro* and neointima formation induced by balloon injury *in vivo*. *J Am Coll Cardiol* 36: pp. 288–93.

Iniesta JA, Gonzalez-Conejero R, Piqueras C, Vicente V, Corral J (2004). Platelet GP IIIa polymorphism HPA-1 (PlA) protects against subarachnoid hemorrhage. *Stroke* 35: pp. 2282–6.

Inoue K, Mineharu Y, Inoue S, Yamada S, Matsuda F, Nozaki K, Takenaka K, Hashimoto N, Koizumi A (2006). Search on chromosome 17 centromere reveals TNFRSF13B as a susceptibility gene for intracranial aneurysm: a preliminary study. *Circulation* 113: pp. 2002–10.

Ito D, Murata M, Watanabe K, Yoshida T, Saito I, Tanahashi N, Fukuuchi Y (2000). C242T polymorphism of NADPH oxidase p22 PHOX gene and ischemic cerebrovascular disease in the Japanese population. *Stroke* 31: pp. 936–9.

Jarinova O, Stewart AF, Roberts R, Wells G, Lau P, Naing T, Buerki C, McLean BW, Cook RC, Parker JS, McPherson R (2009). Functional analysis of the chromosome 9p21.3 coronary artery disease risk locus. *Arterioscler Thromb Vasc Biol* 29: pp. 1671–7.

Jensson O, Gudmundsson G, Arnason A, Blondal H, Petursdottir I, Thorsteinsson L, Grubb A, Lofberg H, Cohen D, Frangione B (1987). Hereditary cystatin C (gamma-trace) amyloid angiopathy of the CNS causing cerebral hemorrhage. *Acta Neurol Scand* 76: pp. 102–14.

Jerrard-Dunne P, Cloud G, Hassan A, Markus HS (2003). Evaluating the genetic component of ischemic stroke subtypes: a family history study. *Stroke* 34: pp. 1364–9.

Johnston SC, Selvin S, Gress DR (1998). The burden, trends, and demographics of mortality from subarachnoid hemorrhage. *Neurology* 50: pp. 1413–8.

Joutel A, Corpechot C, Ducros A, Vahedi K, Chabriat H, Mouton P, Alamowitch S, Domenga V, Cecillion M, Marechal E, Maciazek J, Vayssiere C, Cruaud C, Cabanis EA, Ruchoux MM, Weissenbach J, Bach JF, Bousser MG, Tournier-Lasserve E (1996). Notch3 mutations in CADASIL, a hereditary adult-onset condition causing stroke and dementia. *Nature* 383: pp. 707–10.

Joutel A, Corpechot C, Ducros A, Vahedi K, Chabriat H, Mouton P, Alamowitch S, Domenga V, Cecillion M, Marechal E, Maciazek J, Vayssiere C, Cruaud C, Cabanis EA, Ruchoux MM, Weissenbach J, Bach JF, Bousser MG, Tournier-Lasserve E (1997). Notch3 mutations in cerebral autosomal dominant arteriopathy with subcortical infarcts and leukoencephalopathy (CADASIL), a mendelian condition causing stroke and vascular dementia. *Ann NY Acad Sci* 826: pp. 213–7.

Kaneko Y, Nakayama T, Saito K, Morita A, Sato I, Maruyama A, Soma M, Takahashi T, Sato N (2006). Relationship between the thromboxane A2 receptor gene and susceptibility to cerebral infarction. *Hypertens Res* 29: pp. 665–71.

Kassell NF, Sasaki T, Colohan AR, Nazar G (1985). Cerebral vasospasm following aneurysmal subarachnoid hemorrhage. *Stroke* 16: pp. 562–72.

Kessler C, Spitzer C, Stauske D, Mende S, Stadlmuller J, Walther R, Rettig R (1997). The apolipoprotein E and β-fibrinogen G/A–455 gene polymorphisms are associated with ischemic stroke involving large-vessel disease. *Arterioscler Thromb Vasc Biol* 17: pp. 2880–4.

Khachigian LM, Resnick N, Gimbrone MA Jr, Collins T (1995). Nuclear factor-kappa B interacts functionally with the platelet-derived growth factor B-chain shear-stress response element in vascular endothelial cells exposed to fluid shear stress. *J Clin Invest* 96: pp. 1169–75.

Khurana VG, Sohni YR, Mangrum WI, McClelland RL, O'Kane DJ, Meyer FB, Meissner I (2004). Endothelial nitric oxide synthase gene polymorphisms predict susceptibility to aneurysmal subarachnoid hemorrhage and cerebral vasospasm. *J Cerebr Blood Flow Metab* 24: pp. 291–7.

Kim Y, Lee C (2006). The gene encoding transforming growth factor β1 confers risk of ischemic stroke and vascular dementia. *Stroke* 37: pp. 2843–5.

Kondo S, Hashimoto N, Kikuchi H, Hazama F, Nagata I, Kataoka H (1998). Apoptosis of medial smooth muscle cells in the development of saccular cerebral aneurysms in rats. *Stroke* 29: pp. 181–8.

Kubo M, Hata J, Ninomiya T, Matsuda K, Yonemoto K, Nakano T, Matsushita T, Yamazaki K, Ohnishi Y, Saito S, Kitazono T, Ibayashi S, Sueishi K, Iida M, Nakamura Y, Kiyohara Y (2007). A nonsynonymous SNP in PRKCH (protein kinase C-eta) increases the risk of cerebral infarction. *Nat Genet* 39: pp. 212–7.

Laberge-le Couteulx S, Jung HH, Labauge P, Houtteville JP, Lescoat C, Cecillon M, Marechal E, Joutel A, Bach JF, Tournier-Lasserve E (1999). Truncating mutations in CCM1, encoding KRIT1, cause hereditary cavernous angiomas. *Nat Genet* 23: pp. 189–93.

Lavergne E, Labreuche J, Daoudi M, Debre P, Cambien F, Deterre P, Amarenco P, Combadiere C; GENIC Investigators (2005). Adverse associations between CX3CR1 polymorphisms and risk of cardiovascular or cerebrovascular disease. *Arterioscler Thromb Vasc Biol* 25: pp. 847–53.

Lee BC, Lee HJ, Chung JH (2006). Peroxisome proliferator-activated receptor-γ2 Pro12Ala polymorphism is associated with reduced risk for ischemic stroke with type 2 diabetes. *Neurosci Lett* 410: pp. 141–5.

Lemmens R, Buysschaert I, Geelen V, Fernandez-Cadenas I, Montaner J, Schmidt H, Schmidt R, Attia J, Maguire J, Levi C, Jood K, Blomstrand C, Jern C, Wnuk M, Slowik A, Lambrechts D, Thijs V; International Stroke Genetics Consortium (2010). The association of the 4q25 susceptibility variant for atrial fibrillation with stroke is limited to stroke of cardioembolic etiology. *Stroke* 41: pp. 1850–7.

Leung TH, Ching YP, Yam JW, Wong CM, Yau TO, Jin DY, Ng IO (2005). Deleted in liver cancer 2 (DLC2) suppresses cell transformation by means of inhibition of RhoA activity. *Proc Natl Acad Sci USA* 102: pp. 15207–12.

Levy E, Carman MD, Fernandez-Madrid IJ, Power MD, Lieberburg I, van Duinen SG, Bots GTAM, Luyendijk W, Frangione B (1990). Mutation of the Alzheimer's disease amyloid gene in hereditary cerebral hemorrhage, Dutch type. *Science* 248: pp. 1124–6.

Lianxu C, Hongti J, Changlong Y (2006). NF-kappaBp65-specific siRNA inhibits expression of genes of COX-2, NOS-2 and MMP-9 in rat IL-1beta-induced and TNF-alpha-induced chondrocytes. *Osteoarthr Cartil* 14: pp. 367–76.

Libby P (2002). Inflammation in atherosclerosis. *Nature* 420: pp. 868–74.

Linn FHH, Rinkel GJE, Algra A, van Gijn J (1996). Incidence of subarachnoid hemorrhage: role of region, year and rate of computed tomography: a meta-analysis. *Stroke* 27: pp. 625–9.

Lipton P (1999). Ischemic cell death in brain neurons. *Physiol Rev* 79: pp. 1431–568.

Litchfield WR, Anderson BF, Weiss RJ, Lifton RP, Dluhy RG (1998). Intracranial aneurysm and hemorrhagic stroke in glucocorticoid-remediable aldosteronism. *Hypertension* 31: pp. 445–50.

Lohmussaar E, Gschwendtner A, Mueller JC, Org T, Wichmann E, Hamann G, Meitinger T, Dichgans M (2005). ALOX5AP gene and the PDE4D gene in a central European population of stroke patients. *Stroke* 36: pp. 731–6.

Longstreth WT, Nelson LM, Koepsell TD, van Belle G (1993). Clinical course of spontaneous subarachnoid hemorrhage: a population-based study in King County, Washington. *Neurology* 43: pp. 712–8.

Longstreth WT, Nelson LM, Koepsell TD, van Belle G (1994). Subarachnoid hemorrhage and hormonal factors in women. A population-based case-control study. *Ann Intern Med* 121: pp. 168–73.

Lotta LA, Giusti B, Saracini C, Vestrini A, Volpe M, Rubattu S, Peyvandi F (2010). No association between chromosome 12p13 single nucleotide polymorphisms and early-onset ischemic stroke. *J Thromb Haemost* 8: pp. 1858–60.

Low SK, Takahashi A, Cha PC, Zembutsu H, Kamatani N, Kubo M, Nakamura Y (2012). Genome-wide association study for intracranial aneurysm in Japanese population identifies three candidate susceptible loci and a functional genetic variant at EDNRA. *Hum Mol Genet* (published online).

Lusis AJ (2000). Atherosclerosis. *Nature* 407: pp. 233–41.

Macdonald RL, Pluta RM, Zhang JH (2007). Cerebral vasospasm after subarachnoid hemorrhage: the emerging revolution. *Nat Clin Pract Neurol* 3: pp. 256–63.

Macdonald RL, Weir BK (1991). A review of hemoglobin and the pathogenesis of cerebral vasospasm. *Stroke* 22: pp. 971–82.

Macmillan C, Lach B, Shoubridge EA (1993). Variable distribution of mutant mitochondrial DNAs (tRNA(Leu[3243])) in tissues of symptomatic relatives with MELAS: the role of mitotic segregation. *Neurology* 43: pp. 1586-90.

Majno G, Joris I (1995). Apoptosis, oncosis, and necrosis. An overview of cell death. *Am J Pathol* 146: pp. 3–15.

Margaglione M, Celentano E, Grandone E, Vecchione G, Cappucci G, Giuliani N, Colaizzo D, Panico S, Mancini FP, Di Minno G (1996). Deletion polymorphism in the angiotensin-converting enzyme gene in patients with a history of ischemic stroke. *Arterioscler Thromb Vasc Biol* 16: pp. 304–9.

Markus HS (2011). Stroke genetics. *Hum Mol Genet* 20(R2): pp. R124–31.

Markus HS, Alberts MJ (2006). Update on genetics of stroke and cerebrovascular disease 2005. *Stroke* 37: pp. 288–90.

Maroun FB, Murray GP, Jacob JC, Mangan MA, Faridi M (1986). Familial intracranial aneurysms: report of three families. *Surg Neurol* 25: pp. 85–8.

Masada T, Hua Y, Xi G, Yang GY, Hoff JT, Keep RF (2001). Attenuation of intracerebral hemorrhage and thrombin-induced brain edema by overexpression of interleukin-1 receptor antagonist. *J Neurosurg* 95: pp. 680–6.

Massaro AR, Sacco RL, Mohr JP, Foulkes MA, Tatemichi TK, Price TR, Hier DB, Wolf PA (1991). Clinical discriminators of lobar and deep hemorrhages: the Stroke Data Bank. *Neurology* 41: pp. 1881–5.

Matarin M, Brown WM, Singleton A, Hardy JA, Meschia JF; ISGS investigators (2008). Whole genome analyses suggest ischemic stroke and heart disease share an association with polymorphisms on chromosome 9p21. *Stroke* 39: pp. 1586–9.

Mattson MP (2000). Apoptosis in neurodegenerative disorders. *Nat Rev Mol Cell Biol* 1: pp. 120–9.

Mayne M, Ni W, Yan HJ, Xue M, Johnston JB, Del Bigio MR, Peeling J, Power C (2001). Antisense oligodeoxynucleotide inhibition of tumor necrosis factor-alpha expression is neuroprotective after intracerebral hemorrhage. *Stroke* 32: pp. 240–8.

McPherson R, Pertsemlidis A, Kavaslar N, Stewart A, Roberts R, Cox DR, Hinds DA, Pennacchio LA, Tybjaerg-Hansen A, Folsom AR, Boerwinkle E, Hobbs HH, Cohen JC (2007). A common allele on chromosome 9 associated with coronary heart disease. *Science* 316: pp. 1488–91.

Mehrabian M, Allayee H, Wong J, Shi W, Wang XP, Shaposhnik Z, Funk CD, Lusis AJ (2002). Identification of 5-lipoxygenase as a major gene contributing to atherosclerosis susceptibility in mice. *Circ Res* 91: pp. 120–6.

Meschia JF (2011). Advances in genetics 2010. *Stroke* 42: pp. 285–7.

Meschia JF, Brott TG, Brown RD Jr, Crook R, Worrall BB, Kissela B, Brown WM, Rich SS, Case LD, Evans EW, Hague S, Singleton A, Hardy J (2005). Phosphodiesterase 4D and 5-lipoxygenase activating protein in ischemic stroke. *Ann Neurol* 58: pp. 351–61.

Meschia JF, Worrall BB (2003). New advances in identifying genetic anomalies in stroke-prone probands. *Curr Atheroscler Rep* 5: pp. 317–23.

Montagna P, Gallassi R, Medori R, Govoni E, Zeviani M, Di Mauro S, Lugaresi E, Andermann F (1988). MELAS syndrome: characteristic migrainous and epileptic features and maternal transmission. *Neurology* 38: pp. 751–4.

Morgan L, Cooper J, Montgomery H, Kitchen N, Humphries SE (2006). The interleukin-6 gene −174G→C and −572G→C promoter polymorphisms are related to cerebral aneurysms. *J Neurol Neurosurg Psychiatry* 77: pp. 915–7.

Morgan L, Hawe E, Palmen J, Montgomery H, Humphries SE, Kitchen N (2005). Polymorphism of the heme oxygenase-1 gene and cerebral aneurysms. *Br J Neurosurg* 19: pp. 317–21.

Morgan L, Humphries SE (2005). The genetics of stroke. *Curr Opin Lipidol* 16: pp. 193–9.

Morita A, Nakayama T, Soma M (2006). Association study between C-reactive protein genes and ischemic stroke in Japanese subjects. *Am J Hypertens* 19: pp. 593–600.

Morita H, Kurihara H, Tsubaki S, Sugiyama T, Hamada C, Kurihara Y, Shindo T, Oh-hashi Y, Kitamura K, Yazaki Y (1998). Methylenetetrahydrofolate reductase gene polymorphism and ischemic stroke in Japanese. *Arterioscler Thromb Vasc Biol* 18: pp. 1465–9.

Moriwaki T, Takagi Y, Sadamasa N, Aoki T, Nozaki K, Hashimoto N (2006). Impaired progression of cerebral aneurysms in interleukin-1beta-deficient mice. *Stroke* 37:900–5.

Morooka Y, Waga S (1983). Familial intracranial aneurysms: report of four families. *Surg Neurol* 19: pp. 260–2.

Morrison AC, Doris PA, Folsom AR, Nieto FJ, Boerwinkle E; Atherosclerosis Risk in Communities Study (2001). G-protein β3 subunit and α-adducin polymorphisms and risk of subclinical and clinical stroke. *Stroke* 32: pp. 822–9.

Mun-Bryce S, Wilkerson AC, Papuashvili N, Okada YC (2001). Recurring episodes of spreading depression are spontaneously elicited by an intracerebral hemorrhage in the swine. *Brain Res* 888: pp. 248–55.

Musunuru K, Post WS, Herzog W, Shen H, O'Connell JR, McArdle PF, Ryan KA, Gibson Q, Cheng YC, Clearfield E, Johnson AD, Tofler G, Yang Q, O'Donnell CJ, Becker DM, Yanek LR, Becker LC, Faraday N, Bielak LF, Peyser PA, Shuldiner AR, Mitchell BD (2010). Association of single nucleotide polymorphisms on chromosome 9p21.3 with platelet reactivity: a potential mechanism for increased vascular disease. *Circ Cardiovasc Genet* 3: pp. 445–53.

Naghavi M, Libby P, Falk E, Casscells SW, Litovsky S, Rumberger J, Badimon JJ, Stefanadis C, Moreno P, Pasterkamp G, Fayad Z, Stone PH, Waxman S, Raggi P, Madjid M, Zarrabi A, Burke A, Yuan C, Fitzgerald PJ, Siscovick DS, de Korte CL, Aikawa M, Juhani Airaksinen KE, Assmann G, Becker CR, Chesebro JH, Farb A, Galis ZS, Jackson C, Jang IK, Koenig W, Lodder RA, March K, Demirovic J, Navab M, Priori SG, Rekhter MD, Bahr R, Grundy SM, Mehran R, Colombo A, Boerwinkle E, Ballantyne C, Insull W Jr, Schwartz RS, Vogel R, Serruys PW, Hansson GK, Faxon DP, Kaul S, Drexler H, Greenland P, Muller JE, Virmani

R, Ridker PM, Zipes DP, Shah PK, Willerson JT (2003). From vulnerable plaque to vulnerable patient: a call for new definitions and risk assessment strategies: part I. *Circulation* 108: pp. 1664–72.

Nahed BV, Seker A, Guclu B, Ozturk AK, Finberg K, Hawkins AA, DiLuna ML, State M, Lifton RP, Gunel M (2005). Mapping a Mendelian form of intracranial aneurysm to 1p34.3–p36.13. *Am J Hum Genet* 76: pp. 172–9.

Natowicz M, Kelley RI (1987). Mendelian etiologies of stroke. *Ann Neurol* 22: pp. 175–92.

Neil-Dwyer G, Bartlett JR, Nicholls AC, Narcisi P, Pope FM (1983). Collagen deficiency and ruptured cerebral aneurysms: a clinical and biochemical study. *J Neurosurg* 59: pp. 16–20.

Nevins JR (1998). Toward an understanding of the functional complexity of the E2F and retinoblastoma families. *Cell Growth Differ* 9: pp. 585–93.

Newcombe PJ, Verzilli C, Casas JP, Hingorani AD, Smeeth L, Whittaker JC (2009). Multilocus Bayesian meta-analysis of gene-disease associations. *Am J Hum Genet* 84: pp. 567–80.

Nicoll JA, Burnett C, Love S, Graham DI, Dewar D, Ironside JW, Stewart J, Vinters HV (1997). High frequency of apolipoprotein E epsilon 2 allele in hemorrhage due to cerebral amyloid angiopathy. *Ann Neurol* 41: pp. 716–21.

Nilsson-Ardnor S, Wiklund PG, Lindgren P, Nilsson AK, Janunger T, Escher SA, Hallbeck B, Stegmayr B, Asplund K, Holmberg D (2005). Linkage of ischemic stroke to the PDE4D region on 5q in a Swedish population. *Stroke* 36: pp. 1666–71.

Norrgard O, Angquist KA, Fodstad H, Forsell A, Lindberg M (1987). Intracranial aneurysms and heredity. *Neurosurgery* 20: pp. 236–9.

O'Donnell HC, Rosand J, Knudsen KA, Furie KL, Segal AZ, Chiu RI, Ikeda D, Greenberg SM (2000). Apolipoprotein E genotype and the risk of recurrent lobar intracerebral hemorrhage. *N Engl J Med* 342: pp. 240–5.

Olsson S, Melander O, Jood K, Smith JG, Lövkvist H, Sjögren M, Engström G, Norrving B, Lindgren A, Jern C; International Stroke Genetics Consortium (ISGC) (2011). Genetic variant on chromosome 12p13 does not show association to ischemic stroke in 3 Swedish case-control studies. *Stroke* 42: pp. 214–6.

Onda H, Kasuya H, Yoneyama T, Takakura K, Hori T, Takeda J, Nakajima T, Inoue I (2001). Genomewide-linkage and haplotype-association studies map intracranial aneurysm to chromosome 7q11. *Am J Hum Genet* 69: pp. 804–19.

Opherk C, Peters N, Holtmannspötter M, Gschwendtner A, Müller-Myhsok B, Dichgans M (2006). Heritability of MRI lesion volume in CADASIL: evidence for genetic modifiers. *Stroke* 37: pp. 2684–9.

Orr AW, Sanders JM, Bevard M, Coleman E, Sarembock IJ, Schwartz MA (2005). The subendothelial extracellular matrix modulates NF-kappaB activation by flow: a potential role in atherosclerosis. *J Cell Biol* 169: pp. 191–202.

Ostergaard JR, Oxlund H (1987). Collagen type III deficiency in patients with rupture of intracranial saccular aneurysms. *J Neurosurg* 67: pp. 690–6.

Ostergaard JR, Reske-Nielsen E, Oxlund H (1987). Histological and morphometric observations on the reticular fibers in the arterial beds of patients with ruptured intracranial saccular aneurysms. *Neurosurgery* 20: pp. 554–8.

Ozturk AK, Nahed BV, Bydon M, Bilguvar K, Goksu E, Bademci G, Guclu B, Johnson MH, Amar A, Lifton RP, Gunel M (2006). Molecular genetic analysis of two large kindreds with intracranial aneurysms demonstrates linkage to 11q24-25 and 14q23-31. *Stroke* 37: pp. 1021–7.

Palmer D, Tsoi K, Maurice DH (1998). Synergistic inhibition of vascular smooth muscle cell migration by phosphodiesterase 3 and phosphodiesterase 4 inhibitors. *Circ Res* 82: pp. 852–61.

Palsdottir A, Abrahamson M, Thorsteinsson L, Arnason A, Olafsson I, Grubb A, Jensson O (1988). Mutation in cystatin C gene causes hereditary brain haemorrhage. *Lancet* 2: pp. 603–4.

Pan X, Arauz E, Krzanowski JJ, Fitzpatrick DF, Polson JB (1994). Synergistic interactions between selective pharmacological inhibitors of phosphodiesterase isozyme families PDE III and PDE IV to attenuate proliferation of rat vascular smooth muscle cells. *Biochem Pharmacol* 48: pp. 827–35.

Pavlakis SG, Phillips PC, DiMauro S, De Vivo DC, Rowland LP (1984). Mitochondrial myopathy, encephalopathy, lactic acidosis, and strokelike episodes: a distinctive clinical syndrome. *Ann Neurol* 16: pp. 481–8.

Peters DG, Kassam A, St Jean PL, Yonas H, Ferrell RE (1999). Functional polymorphism in the matrix metalloproteinase-9 promoter as a potential risk factor for intracranial aneurysm. *Stroke* 30: pp. 2612–6.

Ping D, Boekhoudt GH, Rogers EM, Boss JM (1999). Nuclear factorkappa B p65 mediates the assembly and activation of the TNFresponsive element of the murine monocyte chemoattractant-1 gene. *J Immunol* 162: pp. 727–34.

Pluta RM, Hansen-Schwartz J, Dreier J, Vajkoczy P, Macdonald RL, Nishizawa S, Kasuya H, Wellman G, Keller E, Zauner A, Dorsch N, Clark J, Ono S, Kiris T, Leroux P, Zhang JH (2009). Cerebral vasospasm following subarachnoid hemorrhage: time for a new world of thought. *Neurol Res* 31: pp. 151–8.

Pola R, Flex A, Gaetani E, Flore R, Serricchio M, Pola P (2003). Synergistic effect of −174 G/C polymorphism of the interleukin-6 gene promoter and 469 E/K polymorphism of the intercellular adhesion molecule-1 gene in Italian patients with history of ischemic stroke. *Stroke* 34: pp. 881–5.

Pulkes T, Sweeney MG, Hanna MG (2000). Increased risk of stroke in patients with the A12308G polymorphism in mitochondria. *Lancet* 356: pp. 2068–9.

Qureshi AI, Ali Z, Suri MF, Shuaib A, Baker G, Todd K, Guterman LR, Hopkins LN (2003).

Extracellular glutamate and other amino acids in experimental intracerebral hemorrhage: an in vivo microdialysis study. *Crit Care Med* 31: pp. 1482–9.

Qureshi AI, Mendelow AD, Hanley DF (2009). Intracerebral haemorrhage. *Lancet* 373: pp. 1632–44.

Qureshi AI, Tuhrim S, Broderick JP, Batjer HH, Hondo H, Hanley DF (2001). Spontaneous intracerebral hemorrhage. *N Engl J Med* 344: pp. 1450–60.

Raaymakers TWM, Rinkel GJE, Ramos LMP (1998). Initial and follow up screening for aneurysms in familial subarachnoid hemorrhage. *Neurology* 51: pp. 1125–30.

Rabinstein AA, Friedman JA, Weigand SD, McClelland RL, Fulgham JR, Manno EM, Atkinson JL, Wijdicks EF (2004). Predictors of cerebral infarction in aneurysmal subarachnoid hemorrhage. *Stroke* 35: pp. 1862–6.

Reiner AP, Schwartz SM, Frank MB, Longstreth WT Jr, Hindorff LA, Teramura G, Rosendaal FR, Gaur LK, Psaty BM, Siscovick DS (2001). Polymorphisms of coagulation factor XIII subunit A and risk of nonfatal hemorrhagic stroke in young white women. *Stroke* 32: pp. 2580–6.

Richards A, van den Maagdenberg AM, Jen JC, Kavanagh D, Bertram P, Spitzer D, Liszewski MK, Barilla-Labarca ML, Terwindt GM, Kasai Y, McLellan M, Grand MG, Vanmolkot KR, de Vries B, Wan J, Kane MJ, Mamsa H, Schäfer R, Stam AH, Haan J, de Jong PT, Storimans CW, van Schooneveld MJ, Oosterhuis JA, Gschwendter A, Dichgans M, Kotschet KE, Hodgkinson S, Hardy TA, Delatycki MB, Hajj-Ali RA, Kothari PH, Nelson SF, Frants RR, Baloh RW, Ferrari MD, Atkinson JP (2007). C-terminal truncations in human 3′-5′ DNA exonuclease TREX1 cause autosomal dominant retinal vasculopathy with cerebral leukodystrophy. *Nat Genet* 39: pp. 1068–70.

Ridker PM, Hennekens CH, Schmitz C, Stampfer MJ, Lindpaintner K (1997). PIA1/A2 polymorphism of platelet glycoprotein IIIa and risks of myocardial infarction, stroke, and venous thrombosis. *Lancet* 349: pp. 385–8.

Rinkel GJE, Djibuti M, Algra A, van Gijn J (1998). Prevalence and risk of rupture of intracranial aneurysms. *Stroke* 29: pp. 251–6.

Roger VL, Go AS, Lloyd-Jones DM, Benjamin EJ, Berry JD, Borden WB, Bravata DM, Dai S, Ford ES, Fox CS, Fullerton HJ, Gillespie C, Hailpern SM, Heit JA, Howard VJ, Kissela BM, Kittner SJ, Lackland DT, Lichtman JH, Lisabeth LD, Makuc DM, Marcus GM, Marelli A, Matchar DB, Moy CS, Mozaffarian D, Mussolino ME, Nichol G, Paynter NP, Soliman EZ, Sorlie PD, Sotoodehnia N, Turan TN, Virani SS, Wong ND, Woo D, Turner MB; on behalf of the American Heart Association Statistics Committee and Stroke Statistics Subcommittee; On behalf of the American Heart Association Statistics Committee and Stroke Statistics Subcommittee (2012). Heart disease and stroke statistics–2012 update: a report from the American Heart Association. *Circulation* 125: pp. e2–220.

Ronkainen A, Hernesniemi J, Puranen M, Niemitukia L, Vanninen R, Ryynanen M, Kuivaniemi H, Tromp G (1997). Familial intracranial aneurysms. *Lancet* 349: pp. 380–4.

Ronkainen A, Hernesniemi J, Ryynanen M (1993). Familial subarachnoid hemorrhage in east Finland, 1977–1990. *Neurosurgery* 33: pp. 787–96.

Ronkainen A, Niskanen M, Piironen R, Hernesniemi J (1999). Familial subarachnoid hemorrhage. Outcome study. *Stroke* 30: pp. 1099–102.

Roos YB, Pals G, Struycken PM, Rinkel GJ, Limburg M, Pronk JC, van den Berg JS, Luijten JA, Pearson PL, Vermeulen M, Westerveld A (2004). Genome-wide linkage in a large Dutch consanguineous family maps a locus for intracranial aneurysms to chromosome 2p13. *Stroke* 35: pp. 2276–81.

Rosand J, Meschia JF, Singleton AB (2010). Failure to validate association between 12p13 variants and ischemic stroke. *N Engl J Med* 362: pp. 1547–50.

Rosenberg GA (2002). Matrix metalloproteinases in neuroinflammation. *Glia* 39: pp. 279–91.

Rubanyi GM (1991). Endothelium-derived relaxing and contracting factors. *J Cell Biochem* 46: pp. 27–36.

Rubattu S, Stanzione R, Di Angelantonio E, Zanda B, Evangelista A, Tarasi D, Gigante B, Pirisi A, Brunetti E, Volpe M (2004). Atrial natriuretic peptide gene polymorphisms and risk of ischemic stroke in humans. *Stroke* 35: pp. 814–8.

Ruigrok YM, Buskens E, Rinkel GJE (2001). Attributable risk of common and rare determinants of subarachnoid hemorrhage. *Stroke* 32: pp. 1173–5.

Ruigrok YM, Rinkel GJ (2010). From GWAS to the clinic: risk factors for intracranial aneurysms. *Genome Med* 2: p. 61.

Ruigrok YM, Rinkel GJ, van't Slot R, Wolfs M, Tang S, Wijmenga C (2006a). Evidence in favor of the contribution of genes involved in the maintenance of the extracellular matrix of the arterial wall to the development of intracranial aneurysms. *Hum Mol Genet* 15: pp. 3361–8.

Ruigrok YM, Rinkel GJ, Wijmenga C (2005). Genetics of intracranial aneurysms. *Lancet Neurol* 4: pp. 179–89.

Ruigrok YM, Rinkel GJ, Wijmenga C (2006b). The versican gene and the risk of intracranial aneurysms. *Stroke* 37: pp. 2372–4.

Ruigrok YM, Seitz U, Wolterink S, Rinkel GJ, Wijmenga C, Urban Z (2004). Association of polymorphisms and haplotypes in the elastin gene in Dutch patients with sporadic aneurysmal subarachnoid hemorrhage. *Stroke* 35: pp. 2064–8.

Sacco RL (2000). Lobar intracerebral hemorrhage. *N Engl J Med* 342: pp. 276–9.

Sadamasa N, Nozaki K, Hashimoto N (2003). Disruption of gene for inducible nitric oxide synthase reduces progression of cerebral aneurysms. *Stroke* 34: pp. 2980–4.

Saito K, Nakayama T, Sato N, Morita A, Takahashi T, Soma M, Usami R (2006). Haplotypes of the plasminogen activator gene associated with ischemic stroke. *Thromb Haemost* 96: pp. 331–6.

Sakuta R, Goto Y, Horai S, Nonaka I (1993). Mitochondrial DNA mutations at nucleotide positions 3243 and 3271 in mitochondrial myopathy, encephalopathy, lactic acidosis, and stroke-like episodes: a comparative study. *J Neurol Sci* 115: pp. 158–60.

Samani NJ, Erdmann J, Hall AS, Hengstenberg C, Mangino M, Mayer B, Dixon RJ, Meitinger T, Braund P, Wichmann HE, Barrett JH, König IR, Stevens SE, Szymczak S, Tregouet DA, Iles MM, Pahlke F, Pollard H, Lieb W, Cambien F, Fischer M, Ouwehand W, Blankenberg S, Balmforth AJ, Baessler A, Ball SG, Strom TM, Braenne I, Gieger C, Deloukas P, Tobin MD, Ziegler A, Thompson JR, Schunkert H; WTCCC and the Cardiogenics Consortium (2007). Genomewide association analysis of coronary artery disease. *N Engl J Med* 357: pp. 443–53.

Santamaria A, Mateo J, Tirado I, Oliver A, Belvis R, Marti-Fabregas J, Felices R, Soria JM, Souto JC, Fontcuberta J (2004). Homozygosity of the T allele of the 46 C→T polymorphism in the F12 gene is a risk factor for ischemic stroke in the Spanish population. *Stroke* 35: pp. 1795–9.

Schievink WI, Limburg M, Oorthuys JWE, Fleury P, Pope FM (1990). Cerebrovascular disease in Ehlers–Danlos syndrome type IV. *Stroke* 21: pp. 626–32.

Schievink WI, Michels VV, Piepgras DG (1994). Neurovascular manifestations of heritable connective tissue disorders: a review. *Stroke* 25: pp. 889–903.

Schievink WI, Schaid DJ, Michels VV, Piepgras DG (1995). Familial aneurysmal subarachnoid hemorrhage: a community based study. *J Neurosurg* 83: pp. 426–9.

Shearman AM, Cooper JA, Kotwinski PJ, Humphries SE, Mendelsohn ME, Housman DE, Miller GJ (2005). Estrogen receptor α gene variation and the risk of stroke. *Stroke* 36: pp. 2281–2.

Shimo-Nakanishi Y, Urabe T, Hattori N, Watanabe Y, Nagao T, Yokochi M, Hamamoto M, Mizuno Y (2001). Polymorphism of the lipoprotein lipase gene and risk of atherothrombotic cerebral infarction in the Japanese. *Stroke* 32: pp. 1481–6.

Shiraya S, Miwa K, Aoki M, Miyake T, Oishi M, Kataoka K, Ohgi S, Ogihara T, Kaneda Y, Morishita R (2006). Hypertension accelerated experimental abdominal aortic aneurysm through upregulation of nuclear factor kappaB and Ets. *Hypertension* 48: pp. 628–36.

Skirgaudas M, Awad IA, Kim J, Rothbart D, Criscuolo G (1996). Expression of angiogenesis factors and selected vascular wall matrix proteins in intracranial saccular aneurysms. *Neurosurgery* 39: pp. 537–45.

Slowik A, Borratynska A, Pera J, Betlej M, Dziedzic T, Krzyszkowski T, Czepko R, Figlewicz DA, Szczudlik A (2004a). II Genotype of the angiotensin-converting enzyme gene increases the risk for subarachnoid hemorrhage from ruptured aneurysm. *Stroke* 35: pp. 1594–7.

Slowik A, Borratynska A, Turaj W, Pera J, Dziedzic T, Figlewicz DA, Betlej M, Krzyszkowski T, Czepko R, Szczudlik A (2005). α_1-Antichymotrypsin gene (SERPINA3) A/T polymorphism as a risk factor for aneurysmal subarachnoid hemorrhage. *Stroke* 36: pp. 737–40.

Slowik A, Turaj W, Dziedzic T, Haefele A, Pera J, Malecki MT, Glodzik-Sobanska L, Szermer P,

Figlewicz DA, Szczudlik A (2004b). DD genotype of ACE gene is a risk factor for intracerebral hemorrhage. *Neurology* 63: pp. 359–61.

Spanbroek R, Grabner R, Lotzer K, Hildner M, Urbach A, Ruhling K, Moos MP, Kaiser B, Cohnert TU, Wahlers T, Zieske A, Plenz G, Robenek H, Salbach P, Kuhn H, Radmark O, Samuelsson B, Habenicht AJ (2003). Expanding expression of the 5-lipoxygenase pathway within the arterial wall during human atherogenesis. *Proc Natl Acad Sci USA* 100: pp. 1238–43.

Spittell PC, Spittell JA Jr, Joyce JW, Tajik AJ, Edwards WD, Schaff HV, Stanson AW (1993). Clinical features and differential diagnosis of aortic dissection: experience with 236 cases (1980 through 1990). *Mayo Clin Proc* 68: pp. 642–51.

Stehbens WE (1998). Familial intracranial aneurysms: an autopsy study. *Neurosurgery* 43: pp. 1258–9.

Suarez JI, Tarr RW, Selman WR (2006). Aneurysmal subarachnoid hemorrhage. *N Engl J Med* 354: pp. 387–96.

Sun L, Li Z, Zhang H, Ma A, Liao Y, Wang D, Zhao B, Zhu Z, Zhao J, Zhang Z, Wang W, Hui R (2003). Pentanucleotide TTTTA repeat polymorphism of apolipoprotein(a) gene and plasma lipoprotein(a) are associated with ischemic and hemorrhagic stroke in Chinese: a multicenter case-control study in China. *Stroke* 34: pp. 1617–22.

Szolnoki Z, Havasi V, Talian G, Bene J, Komlosi K, Somogyvari F, Kondacs A, Szabo M, Fodor L, Bodor A, Melegh B (2005). Lymphotoxin-α gene 252G allelic variant is a risk factor for large-vessel-associated ischemic stroke. *J Mol Neurosci* 27: pp. 205–11.

Takenaka K, Sakai H, Yamakawa H, Yoshimura S, Kumagai M, Yamakawa H, Nakashima S, Nozawa Y, Sakai N (1999). Polymorphism of the endoglin gene in patients with intracranial saccular aneurysms. *J Neurosurg* 90: pp. 935–8.

Tang J, Liu J, Zhou C, Alexander JS, Nanda A, Granger DN, Zhang JH (2004). MMP-9 deficiency enhances collagenase-induced intracerebral hemorrhage and brain injury in mutant mice. *J Cereb Blood Flow Metab* 24: pp. 1133–45.

Tang J, Liu J, Zhou C, Ostanin D, Grisham MB, Neil Granger D, Zhang JH (2005). Role of NADPH oxidase in the brain injury of intracerebral hemorrhage. *J Neurochem* 94: pp. 1342–50.

ter Berg HW, Bijlsma JB, Veiga Pires JA, Ludwig JW, van der Heiden C, Tulleken CA, Willemse J (1986). Familial association of intracranial aneurysms and multiple congenital anomalies. *Arch Neurol* 43: pp. 30–3.

Teunissen LL, Rinkel GJE, Algra A, van Gijn J (1996). Risk factors for subarachnoid hemorrhage: a systematic review. *Stroke* 27: pp. 544–9.

The International Stroke Genetics Consortium (ISGC); the Wellcome Trust Case Control Consortium 2 (WTCCC2), Bellenguez C, Bevan S, Gschwendtner A, Spencer CC, Burgess AI, Pirinen M, Jackson CA, Traylor M, Strange A, Su Z, Band G, Syme PD, Malik R, Pera J,

Norrving B, Lemmens R, Freeman C, Schanz R, James T, Poole D, Murphy L, Segal H, Cortellini L, Cheng YC, Woo D, Nalls MA, Müller-Myhsok B, Meisinger C, Seedorf U, Ross-Adams H, Boonen S, Wloch-Kopec D, Valant V, Slark J, Furie K, Delavaran H, Langford C, Deloukas P, Edkins S, Hunt S, Gray E, Dronov S, Peltonen L, Gretarsdottir S, Thorleifsson G, Thorsteinsdottir U, Stefansson K, Boncoraglio GB, Parati EA, Attia J, Holliday E, Levi C, Franzosi MG, Goel A, Helgadottir A, Blackwell JM, Bramon E, Brown MA, Casas JP, Corvin A, Duncanson A, Jankowski J, Mathew CG, Palmer CN, Plomin R, Rautanen A, Sawcer SJ, Trembath RC, Viswanathan AC, Wood NW, Worrall BB, Kittner SJ, Mitchell BD, Kissela B, Meschia JF, Thijs V, Lindgren A, Macleod MJ, Slowik A, Walters M, Rosand J, Sharma P, Farrall M, Sudlow CL, Rothwell PM, Dichgans M, Donnelly P, Markus HS (2012). Genome-wide association study identifies a variant in HDAC9 associated with large vessel ischemic stroke. *Nat Genet* 44: pp. 328–33.

The Magnetic Resonance Angiography in Relatives of Patients with Subarachnoid Hemorrhage Study Group (1999). Risks and benefits of screening of intracranial aneurysms in first-degree relatives of patients with sporadic subarachnoid hemorrhage. *N Engl J Med* 341: pp. 1344–50.

Tournier-Lasserve E (2002). New players in the genetics of stroke. *N Engl J Med* 347: pp. 1711–2.

Tournier-Lasserve E, Iba-Zizen MT, Romero N, Bousser MG (1991). Autosomal dominant syndrome with strokelike episodes and leukoencephalopathy. *Stroke* 22: pp. 1297–302.

Tournier-Lasserve E, Joutel A, Melki J, Weissenbach J, Lathrop GM, Chabriat H, Mas J-L, Cabanis E-A, Baudrimont M, Maciazek J, Bach M-A, Bousser M-G (1993). Cerebral autosomal dominant arteriopathy with subcortical infarcts and leukoencephalopathy maps to chromosome 19q12. *Nat Genet* 3: pp. 256–9.

Vajkoczy P, Horn P, Thome C, Munch E, Schmiedek P (2003). Regional cerebral blood flow monitoring in the diagnosis of delayed ischemia following aneurysmal subarachnoid hemorrhage. *J Neurosurg* 98: pp. 1227–34.

van den Berg JS, Limburg M, Pals G, Arwert F, Westerveld A, Hennekam RC, Albrecht KW (1997). Some patients with intracranial aneurysms have a reduced type III/type I collagen ratio. A case-control study. *Neurology* 49: pp. 1546–51.

van der Knaap MS, Smit LM, Barkhof F, Pijnenburg YA, Zweegman S, Niessen HW, Imhof S, Heutink P (2006). Neonatal porencephaly and adult stroke related to mutations in collagen IV A1. *Ann Neurol* 59: pp. 504–11.

Van der Voet M, Olson JM, Kuivaniemi H, Dudek DM, Skunca M, Ronkainen A, Niemela M, Jaaskelainen J, Hernesniemi J, Helin K, Leinonen E, Biswas M, Tromp G (2004). Intracranial aneurysms in Finnish families: confirmation of linkage and refinement of the interval to chromosome 19q13.3. *Am J Hum Genet* 74: pp. 564–71.

Vidal R, Frangione B, Rostagno A, Mead S, Revesz T, Plant G, Ghiso J (1999). A stop-codon mutation in the BRI gene associated with familial British dementia. *Nature* 399: pp. 776–81.

Vila N, Obach V, Revilla M, Oliva R, Chamorro A (2000). α_1-Antichymotrypsin gene polymorphism in patients with stroke. *Stroke* 31: pp. 2103–5.

Visel A, Zhu Y, May D, Afzal V, Gong E, Attanasio C, Blow MJ, Cohen JC, Rubin EM, Pennacchio LA (2010). Targeted deletion of the 9p21 non-coding coronary artery disease risk interval in mice. *Nature* 464: pp. 409–12.

Voetsch B, Benke KS, Damasceno BP, Siqueira LH, Loscalzo J (2002). Paraoxonase 192 Gln→Arg polymorphism: an independent risk factor for nonfatal arterial ischemic stroke among young adults. *Stroke* 33: pp. 1459–64.

Voetsch B, Jin RC, Bierl C, Benke KS, Kenet G, Simioni P, Ottaviano F, Damasceno BP, Annichino-Bizacchi JM, Handy DE, Loscalzo J (2007). Promoter polymorphisms in the plasma glutathione peroxidase (GPx-3) gene: a novel risk factor for arterial ischemic stroke among young adults and children. *Stroke* 38: pp. 41–9.

Wagner KR, Sharp FR, Ardizzone TD, Lu A, Clark JF (2003). Heme and iron metabolism: role in cerebral hemorrhage. *J Cereb Blood Flow Metab* 23: pp. 629–52.

Wang J, Doré S (2007). Inflammation after intracerebral hemorrhage. *J Cereb Blood Flow Metab* 27: pp. 894–908.

Wang J, Fields J, Zhao C, Langer J, Thimmulappa RK, Kensler TW, Yamamoto M, Biswal S, Doré S (2007). Role of Nrf2 in protection against intracerebral hemorrhage injury in mice. *Free Radic Biol Med* 43: pp. 408–14.

Wang PS, Longstreth WT Jr, Koepsell TD (1995). Subarachnoid hemorrhage and family history: a population-based case-control study. *Arch Neurol* 52: pp. 202–4.

Wang X, Mori T, Sumii T, Lo EH (2002). Hemoglobin-induced cytotoxicity in rat cerebral cortical neurons: caspase activation and oxidative stress. *Stroke* 33: pp. 1882–8.

Wang Y, Zhang W, Zhang Y, Yang Y, Sun L, Hu S, Chen J, Zhang C, Zheng Y, Zhen Y, Sun K, Fu C, Yang T, Wang J, Sun J, Wu H, Glasgow WC, Hui R (2006). VKORC1 haplotypes are associated with arterial vascular diseases (stroke, coronary heart disease, and aortic dissection). *Circulation* 113: pp. 1615–21.

Warlow C, Sudlow C, Dennis M, Wardlaw J, Sandercock P (2003). Stroke. *Lancet* 362: pp. 1211–24.

Watnick T, Phakdeekitcharoen B, Johnson A, Gandolph M, Wang M, Briefel G, Klinger KW, Kimberling W, Gabow P, Germino GG (1999). Mutation detection of PKD1 identifies a novel mutation common to three families with aneurysms and/or very-early-onset disease. *Am J Hum Genet* 65: pp. 1561–71.

Wellcome Trust Case Control Consortium (2007). Genome-wide association study of 14,000 cases of seven common diseases and 3,000 shared controls. *Nature* 447: pp. 661–78.

Wiklund PG, Nilsson L, Ardnor SN, Eriksson P, Johansson L, Stegmayr B, Hamsten A, Holmberg D, Asplund K (2005). Plasminogen activator inhibitor-1 4G/5G polymorphism and risk of stroke: replicated findings in two nested case-control studies based on independent cohorts. *Stroke* 36: pp. 1661–5.

Wilhelmsen L, Rosengren A, Eriksson H, Lappas G (2001a). Heart failure in the general population of men—morbidity, risk factors and prognosis. *J Intern Med* 249: pp. 253–61.

Wilhelmsen L, Rosengren A, Lappas G (2001b). Hospitalizations for atrial fibrillation in the general male population: morbidity and risk factors. *J Intern Med* 250: pp. 382–9.

Williams RR, Hunt SC, Heiss G, Province MA, Bensen JT, Higgins M, Chamberlain RM, Ware J, Hopkins PN (2001). Usefulness of cardiovascular family history data for population-based preventive medicine and medical research (the Health Family Tree Study and the NHLBI Family Heart Study). *Am J Cardiol* 87: pp. 129–35.

Woo D, Kaushal R, Chakraborty R, Woo J, Haverbusch M, Sekar P, Kissela B, Pancioli A, Jauch E, Kleindorfer D, Flaherty M, Schneider A, Khatri P, Sauerbeck L, Khoury J, Deka R, Broderick J (2005). Association of apolipoprotein E4 and haplotypes of the apolipoprotein E gene with lobar intracerebral hemorrhage. *Stroke* 36: pp. 1874–9.

Woodruff TM, Thundyil J, Tang SC, Sobey CG, Taylor SM, Arumugam TV (2011). Pathophysiology, treatment, and animal and cellular models of human ischemic stroke. *Mol Neurodeger* 6: p. 11.

Worrall BB, Brott TG, Brown RD Jr, Brown WM, Rich SS, Arepalli S, Wavrant-De Vrieze F, Duckworth J, Singleton AB, Hardy J, Meschia JF (2007). IL1RN VNTR polymorphism in ischemic stroke. Analysis in 3 populations. *Stroke* 38: pp. 1189–96.

Xi G, Keep RF, Hoff JT (2006). Mechanisms of brain injury after intracerebral haemorrhage. *Lancet Neurol* 5: pp. 53–63.

Xia J, Yang QD, Yang QM, Xu HW, Liu YH, Zhang L, Zhou YH, Wu ZG, Cao GF (2004). Apolipoprotein H gene polymorphisms and risk of primary cerebral hemorrhage in a Chinese population. *Cerebrovasc Dis* 17: pp. 197–203.

Xue M, Del Bigio MR (2000). Intracerebral injection of autologous whole blood in rats: time course of inflammation and cell death. *Neurosci Lett* 283: pp. 230–2.

Yamada S, Utsunomiya M, Inoue K, Nozaki K, Inoue S, Takenaka K, Hashimoto N, Koizumi A (2004). Genome-wide scan for Japanese familial intracranial aneurysms: linkage to several chromosomal regions. *Circulation* 14: pp. 3727–33.

Yamada Y, Fuku N, Tanaka M, Aoyagi Y, Sawabe M, Metoki N, Yoshida H, Satoh K, Kato K, Watanabe S, Nozawa Y, Hasegawa A, Kojima T (2009). Identification of CELSR1 as a susceptibility gene for ischemic stroke in Japanese individuals by a genome-wide association study. *Atherosclerosis* 207: pp. 144–9.

Yamada Y, Metoki N, Yoshida H, Satoh K, Ichihara S, Kato K, Kameyama T, Yokoi K, Matsuo H,

Segawa T, Watanabe S, Nozawa Y (2006). Genetic risk for ischemic and hemorrhangic stroke. *Arterioscler Thromb Vasc Biol* 26: pp. 1920–5.

Yamada Y, Metoki N, Yoshida H, Satoh K, Kato K, Hibino T, Yokoi K, Watanabe S, Ichihara S, Aoyagi Y, Yasunaga A, Park H, Tanaka M, Nozawa Y (2008). Genetic factors for ischemic and hemorrhagic stroke in Japanese individuals. *Stroke* 39: pp. 2211–8.

Yasuno K, Bilguvar K, Bijlenga P, Low SK, Krischek B, Auburger G, Simon M, Krex D, Arlier Z, Nayak N, Ruigrok YM, Niemelä M, Tajima A, von und zu Fraunberg M, Dóczi T, Wirjatijasa F, Hata A, Blasco J, Oszvald A, Kasuya H, Zilani G, Schoch B, Singh P, Stüer C, Risselada R, Beck J, Sola T, Ricciardi F, Aromaa A, Illig T, Schreiber S, van Duijn CM, van den Berg LH, Perret C, Proust C, Roder C, Ozturk AK, Gaál E, Berg D, Geisen C, Friedrich CM, Summers P, Frangi AF, State MW, Wichmann HE, Breteler MM, Wijmenga C, Mane S, Peltonen L, Elio V, Sturkenboom MC, Lawford P, Byrne J, Macho J, Sandalcioglu EI, Meyer B, Raabe A, Steinmetz H, Rüfenacht D, Jääskeläinen JE, Hernesniemi J, Rinkel GJ, Zembutsu H, Inoue I, Palotie A, Cambien F, Nakamura Y, Lifton RP, Günel M (2010). Genome-wide association study of intracranial aneurysm identifies three new risk loci. *Nat Genet* 42: pp. 420–5.

Yoneyama T, Kasuya H, Onda H, Akagawa H, Hashiguchi K, Nakajima T, Hori T, Inoue I (2004). Collagen type I α2 (COL1A2) is the susceptible gene for intracranial aneurysms. *Stroke* 35: pp. 443–8.

Yoshimoto Y, Tanaka Y, Hoya K (2001). Acute systemic inflammatory response syndrome in subarachnoid hemorrhage. *Stroke* 32: pp. 1989–93.

Yu Z, Nikolova-Karakashian M, Zhou D, Cheng G, Schuchman EH, Mattson MP (2000). Pivotal role for acidic sphingomyelinase in cerebral ischemia-induced ceramide and cytokine production, and neuronal death. *J Mol Neurosci* 15: pp. 85–97.

Zee RY, Cook NR, Cheng S, Reynolds R, Erlich HA, Lindpaintner K, Ridker PM (2004). Polymorphism in the P-selectin and interleukin-4 genes as determinants of stroke: a population-based, prospective genetic analysis. *Hum Mol Genet* 13: pp. 389–96.

Zhao X, Sun G, Zhang J, Strong R, Song W, Gonzales N, Grotta JC, Aronowski J (2007a). Hematoma resolution as a target for intracerebral hemorrhage treatment: role for peroxisome proliferator-activated receptor gamma in microglia/macrophages. *Ann Neurol* 61: pp. 352–62.

Zhao X, Zhang Y, Strong R, Zhang J, Grotta JC, Aronowski J (2007b). Distinct patterns of intracerebral hemorrhage-induced alterations in NF-kappaB subunit, iNOS, and COX-2 expression. *J Neurochem* 101: pp. 652–63.

Zheng Z, Yenari MA (2004). Post-ischemic inflammation: molecular mechanisms and therapeutic implications. *Neurol Res* 26: pp. 884–92.

Zintzaras E, Rodopoulou P, Sakellaridis N (2009). Variants of the arachidonate 5-lipoxygenase-activating protein (ALOX5AP) gene and risk of stroke: a HuGE gene-disease association review and meta-analysis. *Am J Epidemiol* 169: pp. 523–32.

TITLES OF RELATED INTEREST

Colloquium Series on Genomic and Molecular Medicine

Titles Forthcoming in 2012

Ethical, Legal and Social Issues (ELSI) in Human Genomics
Ruth Chadwick
Cesagen (Centre for Economic and Social Aspects of Genomics), UK

Molecular Basis of Developmental Anomalies of the GI Tract
Charles Shaw-Smith
Royal Devon and Exeter Hospital, Peninsula Medical School, Universities of Exeter and Plymouth, UK

Clinical and Molecular Aspects of Motor Neuron Disease
Pamela J. Shaw
Sheffield Institute for Translational Neuroscience, University of Sheffield, UK

Molecular Basis of Disorders of Porphyrin and Heme Metabolism
Mike Badminton
University Hospital of Wales and School of Medicine, Cardiff University, UK

Molecular Pathology of Multiple Endocrine Neoplasia Syndrome
Rajesh V. Thakker
Nuffield Department of Clinical Medicine, University of Oxford, UK

Molecular Perspectives on Congenital CNS Malformations
Daniela Pilz
Cardiff University School of Medicine, UK

For a full list of published and forthcoming titles:
http://www.morganclaypool.com/page/gmm

Colloquium Series on Integrated Systems Physiology: From Molecule to Function to Disease

Editors

D. Neil Granger, Ph.D., *Boyd Professor and Head of the Department of Molecular and Cellular Physiology at the LSU Health Sciences Center, Shreveport*

Joey P. Granger, Ph.D., *Billy S. Guyton Distinguished Professor, Professor of Physiology and Medicine, Director of the Center for Excellence in Cardiovascular-Renal Research, and Dean of the School of Graduate Studies in the Health Sciences at the University of Mississippi Medical Center*

Physiology is a scientific discipline devoted to understanding the functions of the body. It addresses function at multiple levels, including molecular, cellular, organ, and system. An appreciation of the processes that occur at each level is necessary to understand function in health and the dysfunction associated with disease. Homeostasis and integration are fundamental principles of physiology that account for the relative constancy of organ processes and bodily function even in the face of substantial environmental changes. This constancy results from integrative, cooperative interactions of chemical and electrical signaling processes within and between cells, organs and systems. This eBook series on the broad field of physiology covers the major organ systems from an integrative perspective that addresses the molecular and cellular processes that contribute to homeostasis. Material on pathophysiology is also included throughout the eBooks. The state-of the art treatises were produced by leading experts in the field of physiology. Each eBook includes stand-alone information and is intended to be of value to students, scientists, and clinicians in the biomedical sciences. Since physiological concepts are an ever-changing work-in-progress, each contributor will have the opportunity to make periodic updates of the covered material.

For a full list of published and forthcoming titles:
http://www.morganclaypool.com/toc/isp/1/1

Colloquium Series on Developmental Biology

Editor

Daniel S. Kessler, Ph.D., *Associate Professor of Cell and Developmental Biology, Chair, Developmental, Stem Cell and Regenerative Biology Program of CAMB, University of Pennsylvania School of Medicine*

Developmental biology is in a period of extraordinary discovery and research. This field will have a broad impact on the biomedical sciences in the coming decades. Developmental Biology is interdisciplinary and involves the application of techniques and concepts from genetics, molecular biology, biochemistry, cell biology, and embryology to attack and understand complex developmental mechanisms in plants and animals, from fertilization to aging. Many of the same genes that regulate developmental processes underlie human regulatory gene disorders such as cancer and serve as the genetic basis of common human birth defects. An understanding of fundamental mechanisms of development is providing a basis for the design of gene and cellular therapies for the treatment of many human diseases. Of particular interest is the identification and study of stem cell populations, both natural and induced, which is opening new avenues of research in development, disease, and regenerative medicine. This eBook series is dedicated to providing mechanistic and conceptual insight into the broad field of Developmental Biology. Each eBook is intended to be of value to students, scientists and clinicians in the biomedical sciences.

For a full list of published and forthcoming titles:
http://www.morganclaypool.com/toc/deb/1/1

Colloquium Series on The Developing Brain

Editor
Margaret McCarthy, PhD., *Professor, Department of Physiology; Associate Dean for Graduate Studies; and, Acting Chair, Department of Pharmacology & Experimental Therapeutics, University of Maryland School of Medicine*

The goal of this series is to provide a comprehensive state-of-the art overview of how the brain develops and those processes that affect it. Topics range from the fundamentals of axonal guidance and synaptogenesis prenatally to the influence of hormones, sex, stress, maternal care and injury during the early postnatal period to an additional critical period at puberty. Easily accessible expert reviews combine analyses of detailed cellular mechanisms with interpretations of significance and broader impact of the topic area on the field of neuroscience and the understanding of brain and behavior.

For a list of published and forthcoming titles:
http://www.morganclaypool.com/toc/dbr/1/1

Colloquium Series on Neuropeptides

Editors
Lakshmi Devi, Ph.D., *Professor, Department of Pharmacology and Systems Therapeutics, Associate Dean for Academic Enhancement and Mentoring, Mount Sinai School of Medicine, New York*
Lloyd D. Fricker, Ph.D., *Professor, Department of Molecular Pharmacology, Department of Neuroscience, Albert Einstein College of Medicine, New York*

Communication between cells is essential in all multicellular organisms, and even in many unicellular organisms. A variety of molecules are used for cell-cell signaling, including small molecules, proteins, and peptides. The term 'neuropeptide' refers specifically to peptides that function as neurotransmitters, and includes some peptides that also function in the endocrine system as peptide hormones. Neuropeptides represent the largest group of neurotransmitters, with hundreds of biologically active peptides and dozens of neuropeptide receptors known in mammalian systems, and many more peptides and receptors identified in invertebrate systems. In addition, a large number of peptides have been identified but not yet characterized in terms of function. The known functions of neuropeptides include a variety of physiological and behavioral processes such as feeding and body weight regulation, reproduction, anxiety, depression, pain, reward pathways, social behavior, and memory. This series will present the various neuropeptide systems and other aspects of neuropeptides (such as peptide biosynthesis), with individual volumes contributed by experts in the field.

For a list of published and forthcoming titles:
http://www.morganclaypool.com/toc/npe/1/1

Colloquium Series on The Cell Biology of Medicine

Editor
Joel Pardee, Ph.D. *President, Neural Essence; formerly Associate Professor and Dean of Graduate Research, Weill Cornell School of Medicine*

In order to learn we must be able to remember, and in the world of science and medicine we remember what we envision, not what we hear. It is with this essential precept in mind that we offer the Cell Biology of Medicine series. Each book is written by faculty accomplished in teaching the scientific basis of disease to both graduate and medical students. In this modern age it has become abundantly clear that everyone is vastly interested in how our bodies work and what has gone wrong in disease. It is likewise evident that the only way to understand medicine is to engrave in our mind's eye a clear vision of the biological processes that give us the gift of life. In these lectures, we are dedicated to holding up for the viewer an insight into the biology behind the body. Each lecture demonstrates cell, tissue and organ function in health and disease. And it does so in a visually striking style. Left to its own devices, the mind will quite naturally remember the pictures. Enjoy the show.

For a list of published and forthcoming titles:
http://www.morganclaypool.com/toc/cbm/1/1

Colloquium Series on Stem Cell Biology

Editor
Wenbin Deng, Ph.D., *Cell Biology and Human Anatomy, Institute for Pediatric Regenerative Medicine, School of Medicine, University of California, Davis*

This Series is interested in covering the fundamental mechanisms of stem cell pluripotency and differentiation, and the strategies of translating fundamental developmental insights into discovery of new therapies. The emphasis is on the roles and potential advantages of stem cells in developing, sustaining and restoring tissue after injury or disease. Some of the topics included will be the signaling mechanisms of development and disease; the fundamentals of stem cell growth and differentiation; the utilities of adult (somatic) stem cells, induced pluripotent stem (iPS) cells and human embryonic stem (ES) cells for disease modeling and drug discovery; and the prospects for applying the unique aspects of stem cells for regenerative medicine. We hope this Series will provide the most accessible and current discussions of the key points and concepts in the field, and that students and researchers all over the world will find these in-depth reviews to be useful.

For a list of published and forthcoming titles:
http://www.morganclaypool.com/page/scb

Colloquium Series on The Genetic Basis of Human Disease

Editor
Michael Dean, Ph.D., *Head, Human Genetics Section, Senior Investigator, Laboratory of Experimental Immunology National Cancer Institute (at Frederick)*

This series will explore the genetic basis of human disease, documenting the molecular basis for rare and common; Mendelian and complex conditions. The series will overview the fundamental principles in understanding such as Mendel's laws of inheritance, and genetic mapping through modern examples. In addition current methods (GWAS, genome sequencing) and hot topics (epigenetics, imprinting) will be introduced through examples of specific diseases.

For a full list of published and forthcoming titles:
http://www.morganclaypool.com/page/gbhd

Colloquium Series on Systems Biology and Data Integration

Editors
Aristotelis Tsirigos, Ph.D., *Research Scientist, IBM Computational Biology Center, IBM Research*
Gustavo Stolovitzky, Ph.D., *Manager, Functional Genomics and Systems Biology, IBM Computational Biology Center, IBM Research*

Since the beginning of the 21st century, the development of high-throughput techniques has accelerated discovery in biology. The influx of an unprecedented amount of data presents challenges as well as great opportunities for improving our understanding of living systems. Systems Biology is an interdisciplinary field which, by integrating diverse types of data, such as Genomics, Epigenomics, Proteomics, Metabolomics, etc., aims at modeling biological processes at a systems level - tissue, organ, organism - both in normal function and under stress. This eBook series is dedicated to an in-depth presentation of topics in Systems Biology of research concerning human disease. Each book is written by an expert research scientist who has extensive experience in a particular model system of disease and has demonstrated in his work the value of integrating multiple types of data with potential practical therapeutic applications. Its intended audience is students and scientists in the biomedical sciences who are interested in participating in this fascinating research game of discovery.

For a full list of published and forthcoming titles:
http://www.morganclaypool.com/page/sbdi

Colloquium Series on Molecular Mechanisms in Critical Care Medicine

Editor
Lew Romer, M.D., *Associate Professor of Anesthesiology and Critical Care Medicine, Cell Biology, Biomedical Engineering, and Pediatrics, Center for Cell Dynamics, Johns Hopkins University School of Medicine*

The idea is to provide researchers and critical care fellows with the context and tools to address the burgeoning data stream on the genetic and molecular basis of critical illness. This will be clustered around classical systems including CNS, Respiratory, Cardiovascular, Hepatic, Renal, and Hematologic.

We will explore major themes and categories of current research and practice, such as: to identify genes and molecules involved in control mechanisms for homeostasis and injury response; to identify developmental processes in which these genetic and molecular control mechanisms have a role; and to identify diseases in which these genetic and molecular control mechanisms go awry. The Series will also seek to explain and anticipate mechanisms by which critical care interventions modulate or exacerbate dysfunction of these genes and molecules. Similarly, this Series will seek to describe the role of genotypic variation in explaining the variable phenotypes of critical illness and the interaction between genetic and environmental factors in producing the variable phenotype of critical illness.

For a full list of published and forthcoming titles:
http://www.morganclaypool.com/page/mmccm